The R.A.M.S. Library of Alchemy

I0489318

Volume 12

Chemical Secrets
and
Experiments

by Sir Kenelm Digby

Translated by George Hartman

R.A.M.S. Publishing Company

Chemical Secrets
and
Experiments

by Sir Kenelm Digby

Produced by

Restorers of Alchemical Manuscripts Society
1983

R.A.M.S. Publishing Company

R.A.M.S. Publishing Company
117 Rutherford Lane
Stuarts Draft VA 24477

The R.A.M.S. Library of Alchemy, Volume 12:
Chemical Secrets and Experiments
Copyright © 2016 R.A.M.S. Publishing Company

All rights reserved. No part of this publication may be
reproduced or transmitted in any form or by any means, electronic or
mechanical, including but not limited to any information storage and
retrieval system, without written permission from R.A.M.S Publishing
Company. Reviewers may quote brief passages.

First Edition 2016

ISBN-13 **978-1523953462**
ISBN-10 **1523953462**

Image Processing by Philip N. Wheeler

This book is sold for informational purposes only. Neither the
publisher nor the editor shall be held accountable for the use or misuse
of the information in this book.

Printed in the United States of America

Table of Contents

11

Dedicated to Hans W. Nintzel,

American Alchemist

and

Founder of the

Restorers of Alchemical Manuscripts Society

(R.A.M.S.)

Disclaimer

Liability: The publisher does not warrant or assume any legal liability or responsibility for the accuracy, completeness, or usefulness of any information, apparatus, product, or process disclosed. The publisher makes no representation as to the accuracy or completeness of the contents of this book and specifically disclaims any implied warranty of merchantability or fitness for a particular purpose. No warranty may be created or extended by written sales materials or sales representatives. You should obtain professional consultation where appropriate. The publisher shall not be liable for any loss of profit or other commercial or personal damages, including but not limited to special, incidental, consequential, or other damages.

Sir KENELM DIGBY

Chemical
Secrets
and
EXPERIMENTS

Translated by:

George Hartman
LONDON 1608

PRODUCED BY:

RAMS

1983

19

Sir Kenelm Digby

A Choice

COLLECTION

of rare

Chemical Secrets

and

EXPERIMENTS

in

Philosophy

As Also

RARE AND UNHEARD OF *MEDICINES, MENSTRUUMS and ALKAHESTS.*

with the True Secret of Volatilizing the fixt Salt of

TARTAR

collected

And experimented by the Honourable and truly learned Sir

KENELM DIGBY

Kt. & Chancellor to Her Majesty the Queen-Mother, Eng.

Hitherto kept secret since his decease, but now Published
for the good and benefit of the Publick by: GEORGE HARTMAN.

22

Introduction

Philip N. Wheeler

Sir Kenelm Digby (1603 - 1665) played a lively part in the chemical and alchemical studies of his time. He was an English courtier and diplomat, and a highly reputed natural philosopher.

Digby is the inventor of the modern wine bottle. His manufacturing technique involved a coal furnace, made hotter than usual by the inclusion of a wind tunnel, and a higher ratio of sand to potash and lime than was customary. Digby's technique produced wine bottles which were stronger and more stable than most existing wine bottles, and due to their translucent green or brown color they protected the contents from light.

His *Chemical Secrets* gives us considerable insight into the studies of science in his time.[1] Throughout the text, the translator (George Hartman) provides commentary on the processes being described.

Most of the symbols in the text were copied from the original R.A.M.S. manuscript produced by Hans W. Nintzel. You may want to obtain a copy of **Alchemical Symbols**, R.A.M.S. Library of Alchemy Volume 21, to assist in their interpretation. I also had access to an original printed copy of the 1682 edition, which I used to verify the R.A.M.S. manuscript's content.

Chymical secrets and rare experiments in physick & philosophy with figures collected and experimented / by the Honourable and learned Sir Kenelm Digby ... Digby, Kenelm, Sir, 1603-1665. London: Printed for Will Cooper, 1683.

---

[1] "The Foundations of Newton's Alchemy" by B. J. T. Dobbs

An Explication of the Characters, which are used in this Book.

☉ Gold. Sun.

☽ Silver. Moon.

♂ Iron. Mars.

☿ Mercury.

♃ Tin. Jupiter.

♀ Venus. Copper.

♄ Lead. Saturn.

♁ Antimony.

✳ Sal armoniac. Sal ammoniac.

A. F. Aqua Fortis.

A. R. Aqua Regis.

S. V. Spirit of Wine.

🝞 Sublimate.

🝟 Precipitate.

ááá Amalgama.

▽ Water.

△ Fire.

To The Right Honourable Robert,

Lord Paston, Baron of Paston,
Viscount and Earl of Tarmouth

My Lord:

It is not my intention, nor indeed my Talent, to Celebrate those Excellent Virtues, which shine so eminently bright in your Lordship: For they are Themes only fit to be treated on by the strongest Pen, and their Native Worth and Resplendency are their own sufficient Panegyricks.

Yet, such is the Veneration I have for the Excellent Qualities and Endowments of your Noble Mind, and those Heroic inclinations, that move your Honour to make such exact diligent, and curious search into all the Secrets and Mysteries of Nature, and encourage all others that Labor therein, that I cannot forbear being so vain as to publish my resentments thereof to the World. These Reasons, My Lord, together with the consideration of your inbred Candour and Generosity, encouraged me to the boldness of committing this small Treatise to your Honours Patronage and Protection: To whom I know it will be welcome as containing in it the Choice Observations both in Physics and Chemistry, of that Famous Man, and great Privy Counsellor of Nature, Sir K. D., A Name, My Lord, that has peculiar Charms with it, to recommend all that are under its great shadow, to the value and consideration of all the Diligent, the Learned and the Honourable: So great a Person (may I assume this vanity to say so) I had the Honour and Happiness for several years to serve, beyond the Seas, as well as in England, and to attend on him more particularly in the Production of many of his incomparable Experiments, and so to continue to his dying day; when he left with me those Choice things contained in this little Treatise.

And since I fear they suffer diminution of their Worth and Beauty, by passing through my mean Hands, and weak Managery, I thought I could make no better atonement, then by recommending them to receive Recruits, and Reinforcements from the Splendor and imminence of your Illustrious Name. To that end therefore I take the Boldness to prostrate them at your Honours Feet, where also in all humility lies

Your Honors Most Obedient,
and Most Devoted Servant,
George Hartman.

To The Reader

Courteous Reader,

This Treatise contains some of the Choicest Collections of the Famous Sir Kenelm Digby (some whereof have been wrought by his own hands, some communicated to him by Learned Men of all Nations) in praise whereof no more need be said, but, that they are his; either of his production, or of his approbation. I shall, therefore, omitting other Artifices and Insinuations, only satisfy the Reader with all the clearness and ingenuity I can, how I came by them; and thereby I question not, but I shall very successively recommend this Collection of them to all Ingenious Lovers of Art, whose Ears the Fame of the Worthy Author has reached.

To this End, I must acquaint him, that I had the Honour for several years to wait upon Sir Kenelm, and to have several of his Manuscripts in my Custody, more particularly this with others was committed to my Charge, when my Worthy Master intended a Journey to France for his Health's sake and to settle his Affairs there: And he had progressed in his designed Journey as far as Cittenburn, when a violent Distemper forced him back to his own House in Covent Garden; and in three days after his return, left the Learned World to lament the loss of so great a Man. And here besides his incomparable self, his Friends and Country lost the benefit of his Famous Library he had in France (which for want of his being Naturalized) fell into the French Kings hands, who bestowed it upon a gentleman, and it was sold (as I was credibly informed) for ten thousand Crowns. In which no doubt were manuscripts of his own, of very great worth and rich value, and might have obliged the world, had they fallen into the hands of Generous and Communicative Men.

But it was my happiness to have, among some other Manuscripts of his, the sole Propriety of this Choice Manual, which contains rare and profitable Secrets in Philosophy and Chemistry, delivered with more perspicuity and plainness than is yet to be found in any Chemical Book: Yea, such that any understanding Reader may with great facility be conducted as with an *Ariadnean* Line into the most intricate and hitherto fatiguing Arcana of Chemistry. Here are the true Menstruums and Alkahests, and that hitherto hidden Secret of Volatizing the fixed Salt of Tartar without any Heterogeneous Substance, being the true Vegetable Menstruum; with many other rare and unheard of Medicines, some whereof I had a mind to reserve for myself, and not to Publish them during my Life, because of the great Experience, and the good success I have had with them in desperate cases; but *Impium est sacere quae promulgata essent, multorum misere decumbentium, in levamen extarent.* 'Tis Impious and unchristian to forbear the Publication of those things, which being rendered Public, will effectually redound to the Advantage and Comfort of miserable Man.

I have translated most of these Secrets out of the Latin, French, German and Italian Tongues. And if I have committed any indecencies against the Idiom of the English Speech, I hope the Candid Reader will forgive a Foreigner. As for any Ornaments and Embellishments of Language, as the Work requires no such a Painting, so the Publishers Foreign Style and mean Talents are not able to afford it.

I have no more to acquaint the Reader with, but that these Secrets had been Communicated to him long before, but that I have been most part of my time since my Excellent Masters Decease, abroad: However, I hope they will be now kindly entertained. For it is the height of my Ambition to make the Memory of my incomparable Master to Live, who was my private, and the World's Public Benefactor, which can nowhere do with such Advantage as in his Learned Works, for thereby being dead, he yet speaks and instructs.

And though it is no addition to his Glorious Name, yet for the Wit and smartness of the thing, and the Readers diversion, I shall conclude here with that Elegant Epitaph made on him by the ingenious Dr. Farrar, which is as follows.

An Epitaph

UPON

The Honorable and Truly Nobel

Sir Kenelm Digby, Kt.,

Chancellor to Her

Majesty

THE

QUEEN MOTHER

Under this Tomb the Matchless Digby lies;
Digby the Great, the Valiant and the Wise:
This Ages Wonder for his Noble Parts,
Skilled in Six Tongues, and Learn'd in all the Arts.

Born on the Day He Dy'd, Th' Eleventh of June,
And that Day Bravely Fought at Scandaroun:
'Tis Rare, that one and the same Day should be
His Day of Birth, of Death and Victory,

R.F.

CHEMICAL SECRETS

How to Fix ☽ into ☉

by ♀ and ☿ Precipitate.

Having written so many processes, and made so many trials, and heard so many discourses of learned men upon this subject, I will give you an account of an easy method that I have resolved upon for accomplishing this work. Namely: That all imperfect metals and common ☿ may be transmuted into ☉ by one and the same method; to wit, by maturation and coction, and not by generation; for that which is generated, is no more that which it was before it was generated: And that which is corrupted, is no more that which it was before it was corrupted.

But the baser metals, after they are transmuted into ☽ or ☉ are still metals nevertheless as they were before, and the transmutation of their kind is done by changing their accidental form, not their substantial, the perfection whereof is Maturity; for by maturation the metal is brought to a higher degree of perfection.

Now, imperfect metals are maturated by external heat, which digested their crude humidity; yea, ☉ itself may be further perfected, and exalted in colour, as when the Stone is made of it, it will communicate this Maturity to imperfect metals.

And common ☿ is extracted out of metals three ways; namely, by cementation and digestion, by fermentation, and by tincture. As for cementation, that concerns only the digestion of ☽ into ☉ but not the other too imperfect metals, nor ☿ either, which is crude, and too much alienated from the Maturity of ☉.

By cementation, the humidity of ☽ is brought to Maturity. There are several sorts of cementations, namely by Salts, Alloms, Vitriols, and Metalline waters. But often-times instead of digesting ☽, they burn it; so that this way of making ☉ is with more loss than profit.

But there is no better way than ☿ and red Precipitate, which I have learned by the afore-mentioned operations.

My Process is Such.

Take ℥ij. of ☽, make an Amalgam of it with Mercury by A. F. as you know, wash this Amalgam very well in several waters, then press out so much ☿, that there remain but ℥iiij. of it with the ☽ which made ℥vj. in all. Add to ℥vj. of good red

Precipitate, and grind all to an impalpable Powder; which put into a Matrass, and digest it with a gentle heat in Sand, so that the ☿ does not sublime, but that it may calcine the ☽, and leave it in powder, for if you give too great a heat, it will reduce the ☽ into a body.

After three days digestion, take out your powder; and grind it as before, so that if there is still any quick ☿, it may be mortified. Digest it again as before, and with the same degree of heat for three days more; then take it out and grind it again, then digest it only for two days by four degrees of heat, which you must change every two hours, to the end, that by the last degree of heat all the ☿ and ☿ precipitate may precipitate the powder of ☽, which will begin to grow white.

Reiterate the addition of ☿ and ☿ precipitate in the same quantity as before. Digest two days more by four degrees of heat, as before, and the powder will grow perfectly white.

Then by the same operation reiterated it will begin to be of a Citrine colour, and fixed.

And thus by reiterating the digestions, you may give it what degree of this colour you please; for the more often you digest it with the said ☿ and ☿ precipitate, being separated from it again by the last degree of heat, the more the powder will be of a Citrine Colour.

Then melt your powder with Borax, and you shall have ☉ at 24 Carats, without diminution of the first weight of the ☽ which will be rather increased. All may be accomplished in the space of One and Twenty days.

A Work With ☉ and ☿

That Monsieur Dandre
Helped to Work in Piedmont, in Great Quantity:
Given me by Him, June 1660.

Monsieur Dandre said, he wrought thus: Make carefully an amalgama of ℥j. of ☉ in Calx, with 7 or 8 of purified ☿, then squeeze out so much ☿ that there remain ℥iij. of ☿, and so there is ℥iv. in the Globe: To this add ℥j. of Sulphur-vive, which is clear and transparent in pieces, (in Italy, where he wrought this) and grind all very well together, (in this consists the main part of the Secret, for at every time you are to employ three or four hours in grinding) then put the matter into a Matrass, and give a gentle heat, the Glass unstopped, till the moisture and smoke are exhaled out. Then let the fire go out, and when the Matrass is cold, seal it hermetically, and set it to sublime by degrees of Fire, till all that will, is sublimed, which will be done in 20 or 24 hours. Then the Vessel is cold, break it, and take out the Matter, and grind all together a long time, both that which is sublimed, and that which is in the bottom, adding ℥j. of new Sulphur-vive, then sublime in the same method as before; repeat this seven times at the least, adding ℥j. of Sulphur-vive every time, and the matter will become a brown yellowish reddish powder, which will be very fusible, and even in the grinding it will relent, as though it grew moist: You will have ℥iv. of fixed Matter, which project (in parcels) upon ℥x. of ☽ in good fustion[2], then put it to the Coppel, and separating ▽ and you shall have ℥iv. of pure ☉.

You can work with but ℥j. of ☉ in one glass, but you may put 50 or more glasses in one Sand Furnace with a large bason of Copper in Sand.

---

[2] This is the exact word in the R.A.M.S. manuscript. It also appears in the book printed in 1682. -pnw

Some Observations about the Work
of Monsieur Dandre.

The operation was made in an Athanor, with the Registers at the end, the hole through which the heat was communicated, was about the bigness of a brick, the plate which held the Sand was of Iron, and contained 32 matrasses, sixteen on every side: The Tower was in the middle, wherein the coals sunk down by degrees. They did not mix the powder with wax, or anything else in projecting but only wrapped up in paper, it did enter, and disappeared immediately without smoking. The Matrass ought to have 2 third parts empty.

The Amalgama was made thus: They took ℥ix. of ☿ and heated it in a Crucible, until it began to smoke, then they set it upon hot Ashes, and cast therein ℥j. of Ducats cut in small pieces, and made hot in a crucible, then stirred it until the ☉ was swallowed up by the ☿: Then took it off, and let it cool. They did not wash the said Amalgama: They used common ☿ only mixed with Quick-lime, and then Distilled in a Retort.

The Sulphur was transparent and yellow like Amber, in pieces, and to be had at Carmagnole, Turin, Cony, Mondevic, Saluce, Genes: It is a

Sulphur-vive, cost four, five, or six pence a pound. The Sand they used was round River-sand, and the Matrass was never red in the Sand: They never put above ℥j. of ☉ in one Matrass, nor ever above ℥x. of ☽ at every projection.

(Hartman) These observations were communicated to Sir Kenelm by Abbot Boucaud, but the Process was written by Sir Kenelm himself from Monsieur Dandre's Mouth.

Monsieur van Outer's Secret.

Van Outer was a Physician from Brussels. This section deals with ☉ and Butter of Antimony.

Take equal parts of ♁ mineral, and ☿ sublimate, and a little Sal armoniac, make butter thereof: Draw the spirit from this butter, which rectify again. (Note, that this butter, being exposed to the Air, draws from the same what it needs in an hour's time, and thereby is much increased in quantity: That which it draws is the hidden food of the Life of Man, and all the Beings in the World. And this butter is the true magnet which draws it in its purity.) Then put this spirit into a glass cucurbite, of a convenient bigness, fit a head to it with Limbec and Receiver; lute well all the junctures, and put it thus to putrefy in Ashes

for two months, in which time the matter will become
as red as Blood, and afterwards very black, sticking
to the sides of the vessel like glutinous Soot, and
the Ethereal Spirit ascends and passes into the
Recipient in form of a Spirit, and in a Body of
fusible salt, whereof you must also draw the spirit,
and separate them by distillation with a very gentle
fire, until you see a red and sparkling fire upon
the matter, which is a sign of its Maturity, and
that you have obtained the Philosophical ☿, which is
the true universal dissolvent, then let it cool.
That which rests in the bottom of the cucurbite, is
the *Terra damnata*.

Take ℥vj. of this Menstruum, and put it upon
℥j. of ☉ in very thin plates, which will speedily
be dissolved, and they will unite intimately, as
being of the same nature. You must take great care
that you lose nothing of the spirits; it must be
done in a Matrass with a glass stopper, exactly
fitted; and being well sealed and luted, digest it
with a Lamp fire, with a very gentle heat in the
beginning. After fifty days digestion, you must feed
and imbibe your matter with the said Menstruum,
whereof you must have store, for to multiply your
work. So soon as you have put in the said dissolvent, you must stop it again immediately, and seal
it as before, then digest fifty days more, the heat

a little increased; which time being expired, you must again feed your matter with the Virginal Milk a little more than the first time, continuing the digestion, the heat a little stronger. Reiterate the Imbibition seven times, and your matter will become more vigorous, and will be able to bear stronger food from time to time, and to bear stronger heat, which nevertheless must not be hastened, but well governed, after the example of the operation of the Rays of the Sun in the Spring and Summer, for the nourishment and maturation of vegetables. But you must observe, that at the two last imbibitions, there must be but 35 days distance from the one to the other instead of 50 as before. At the five first imbibitions you shall see from time to time the wonderful effects of Nature, by the internal virtue of the matter, and by all the signs written in Flammel, Le Tourbe, Le Rosaire, or Jubilation of the Soul, and in all those Authors that have possessed this rare Knowledge, which will appear infallibly to the proportion whereof you must increase the fire, and that is left to the discretion of the operator. You must observe, that as the matter multiplies in virtue and quantity at each imbibition, and always more and more, it might become so fusible, that at last it might penetrate the glass; so that if you judge convenient, you need not imbibe quite seven times, that you may run no hazard; for you may afterwards multiply the powder in the same manner, and carry it ad infinitum. And to perform all this, there need be no more than about nine months' time, and without much trouble or care.

The Multiplication of the Powder.

Take ℥j. of the powder to ℥iiij. of ☉, melt them together, and leave them until all are reduced to a powder, which will be done in three days at the most; and thus you may carry it ad infinitum, and that which is made thus, has the same virtue as the first.

The Projection.

To project upon ☿ you must heat it in a Crucible, until it cast a black smoke, then cast one grain of the said powder upon ten or twelve ounces of ♀. And projecting upon other metals, they must be in fusion, and they will render in proportion according as they abound in ☿.

A Considerable Work with ☉ and ☿.

Take ℥viij. of ☉, melt it in a crucible with three times as much Tin-glass, mix them well together, then cast it out, and beat it into as small pieces as you can: Take three times the weight of your mixture of good sublimate, which put in the bottom of a large cucurbite, and upon that put the said mixture; set the cucurbite in an Earthen pot, which put into an Iron pot with Sand; fit a head with a Limbeck and Receiver to it, lute all well, and give a gentle heat at the beginning for two hours; then increase the heat by degrees, at last a very violent fire of reverberation, during eight hours; then let it cool, and open the vessel, and you shall find your Tin-glass in the receiver in the form of Crystals, with the sublimate, and the ☉ will remain in the bottom of the cucurbite, in the form of light dry flowers, very fair to behold, and will be much opened and attenuated.

Dissolve this ☉ in eight parts of A. R. Distill it, and put the same quantity of new A. R. upon it, and distill it off as before. Repeat this three times, at the third time the ☉ will be so opened, that it will ascend with the water, and stick to the sides of the head of the Alembick; so the same will seem to be full of golden Stars.

41

Dissolve this ☉ again in eight parts of A. R. Dissolve also by itself twelve Marcs[3] of ☿ In A. F. Put these two dissolutions together, and let them stand to settle 24 hours, the ☉ and ☿ will be precipitated indistinguishable, in the form of a black sponge, and will be essentially and radically united.

Distill off the water to dryness, you will find at the bottom a gray powder, which take out, and put it into a Matrass, and pour upon it good Oil of Vitriol, so much as may cover it the breadth of four fingers. Seal it hermetically, and digest for twenty days. Then open the Matrass, and let the humidity exhale by a strong heat in Sand: Break the glass, and grind the matter with a little Borax, then melt it, and you shall have at least eleven Marcs of ☉.

Monsieur Carrier gave this work to his Uncle, Monsieur Ferrier, having had it from an intimate Friend of his, who had arrived to great wealth by it.

(Hartman) The said Monsieur Ferrier did communicate this Process to Sir Kenelm at Paris, 1660, when he returned from Germany, at the time of the King's Happy Restauration[4].

---

[3] A unit of weight measure: traditionally a half pound weight, usually divided into 8 ounces or 16 Lots. However, the 1682 printed text says that a Marc is 10 ounces. -pnw
[4] Restauration is French for restoration. -pnw

A Work copied out of the Original of
Monsieur De La Violette's Own Hand,
whereof he made Great Account.

Take ℥iiij. of the purest and finest ♃ , and ℥viij. Spanish ☿ purified with salt and vinegar, make an Amalgama. Then take red Minimum and Aesustum of each ℥iiij. Danzick Vitriol lbj. reduced to half a lb. by calcination, grind and mix these all well together, and put them into a retort coated, and pour upon it one pound and a half of the following A.F.

Take Vitriol two pounds, which reduce to one lb. by calcination, which put into a retort, and pour upon it a good A. F. made of Vitriol and Nitre, Distill it S. A. and you shall have an A. F. fit for this work, which having poured upon the said matter, distill it off, and it will be very ponderous. Break the retort (being cold) and you will find on the sides of it, and upon the Caput Mortuum, a very red and ponderous sublimate, which take off.

Take the half of the Caput Mortuum, and as much of Bay Salt decrepitated, reduce all to a fine powder with the said sublimate, and then put all into a new retort, and pour upon it the distilled A. F. Distill it as before, and the said A. F. will come off very red, and the sublimate will be more red, and more ponderous than before, and will rise very high at this time. Keep this water very carefully, break the retort, and take both the feces and sublimate, and reduce it to powder, and sublime it by itself without A. F. and the sublimate will mount but upon the surface of the feces, which separate, and it will have acquired more redness, and will be almost fixed. Put this sublimate into the said A. F. and it will dissolve it speedily: Distill or evaporate the A. F. in sand, and the sublimate will remain in the bottom like a deep-red Oil. Put into this Oil $\mathfrak{Z}$iij. of the fixed Sulphur of Vitriol, made according to Art; put it into a Matrass with a short neck, and digest in sand, until all the moisture is exhaled.

Then take an Amalgam made with one part of ☉ and two parts of ☽ calcined with salt, and four parts of Spanish ☿ (washed with salt and vinegar;) then squeeze out so much ☿ as you can from the Amalgama, then wash and dry this Amalgama, and pour upon it by little and little of the above-said A. F. Let it stand half an hour, then pour on more of the water as before, and you will see the Amalgama dissolve visibly, and will be reduced to a very red powder.

Note, that once in half an hour you must pour on some of the said water, and all will be done in less than half a day. Digest it half a day longer in sand; then break the vessel, take out this precipitate, and melt it with a little Borax, and you shall have ☉ at 24 Carats.

Note, that if you take equal parts of ☉ and ☽ to your Amalgama, you shall have increase yet forty or fifty percent more.

Snyder's Secret,

as he gave it to me himself, the 22 of July, 1664.

Take Nitre eight parts, Sulphur four parts, and Tartar two parts: Reduce all into a fine powder, and mix them well. Then melt one part of pure ☉ and three parts of purified Regulus of Antimony in a crucible; then add to them three parts or more of the said powder, let it stand in the fire until you see a light skin upon it, then pour it into an Antimony-horn. Take the Regulus in the bottom of the Horn, and melt it again, and cast more of the said powder upon it: Repeat this so often until all the Regulus be consumed; dissolve all the scums of the said Regulus, and make a layer thereof, which filter, and precipitate with an Acid, which edulcorate; edulcorate also the feces which remained

45

in the filter, put these things edulcorated together, with half the weight of flowers of sulphur, and calcine them well: Then draw the salt from it with distilled vinegar (which will be a golden salt) draw as much of the said salt from it as you can.

Take one part of this salt, and two or three parts of good butter of Antimony well rectified, mix them well in a Matrass, one part filled, and the other two parts empty: Seal it Hermetically, and digest it with a gentle heat; it will grow black and putrefy in the space of three days; continue the digestion until the powder is fixed.

Observations from Another Learned Man,

**with whom Sir Kenelm did confer,
upon his return from Bristol, concerning the said
Snyder's Work, who said thus.**

This operation may be abbreviated, in fermenting it with ☉ as follows: Make a Spiritus Regulus of ♂, as you know, which is precipitated butter of ♂ and ☿, adding to them soap and salt of Tartar. Take of this Spirituous Reg.three parts, and one part of ☉ melt them together, and cast it by little and little into the Sulphurous Salt Enixe, &

totus solvitur, effunde, solve, filtra, precipita totam materiam in Sulphur pulcherimum: Reverberate this Sulphur with flowers of Sulphur if you will, dissolve it again, and precipitate; draw the salt from this Sulphur with distilled Vinegar; add to this salt or Golden Vitriol, three times its weight of Butter of ♂; digest them together (*domo cossent colores*.) You may multiply the work in quality, in dissolving the powder in Salt Enixe, and precipitating often: And you may multiply it in quantity, in mixing it with new Butter of ♂, wherein you have dissolved the said salt, or Golden Vitriol. Note, That this work will be more excellent if it is done with ☿ of ♂, and Spirituous Reg. It may be also abbreviated by purifying very well the Butter of ♂. Note, that this works in a mineral water, which is coagulated by its own Sulphur. Note also, that if you take the Golden Sulphur without Reg. the work will be yet shorter. Note, that in the multiplication, if the powder is only dissolved in Butter of ♂, the operation will be shorter.

A Great Secret of Mr. Snyder's Powder.

Dissolve ☉ in Sal Enixe, and exalt it with Sulphur of ♂, then cast in *conum, in salem rubicundum* (see that no Coals fall in.) Keep the salt so long in the fire, that it remains fusible. Grind it, and let it melt in a Matrass; add a grain or two of the powder, let all melt in a strong fire twelve to twenty hours, and this powder will be multiplied; pour out, dissolve, and filter, put therein ☽ and ☿, they will be transmuted into fine ☉.

   Or, precipitate the Liquor with salt into a Golden Sulphur, which digest longer with Butter of ♁. Or, preserve the Sulphur, and ferment it again with dissolved ☉, as is said, in Salt Enixe, and in a Matrass, that the powder may go *ad infinitum*.

Matthews his Work.

   Take Common Cinaber ℥xij. Crystals of ♂ ℥ij. Common ☿ precipitate, made by A. F. and reverberated until it is red, ℥j. Oil of Vitriol ℥xv. First, reduce the three hard ingredients into a most fine powder; then grind it upon a Marble stone with a little of the Oil of Vitriol, adding the said Oil by little and little, until it becomes like Pap; which put into a low cucurbite (taking care that it does not touch the sides of the said cucurbite, because it would endanger it to break) and put upon it the rest of the Oil of Vitriol, and stir the matter well with a stick of glass (which must be massy and not hollow) that all may be well mixed together; digest it with a gentle heat for eight days, so that nothing may go over through the Limbeck: Then distill as much as you can of the Oil of Vitriol, and take the matter out of the cucurbite, and grind it again; put the distilled Oil upon it again, and distill as before, without digesting it; repeat this fourteen or sixteen times. At last, distill as much of the Oil as possibly you can; and that the remaining matter may be thick, and conveniently handled, put into it ℥v. or vj. of filings of ☽. Then melt twenty ounces of ☽, and project your matter upon the same (being in fusion) in fifteen or twenty parcels, staying every time, before you project, until that which you projected be well entered and incorporated with the ☽, and that it be very clear. After all is projected, leave it in good fusion for an hour or two; then put it to Coppel, and afterwards to a Separating ▽ and you shall have about ℥j ß. of pure ☉.

The Crystals of ♂ are made thus:

Upon filings of ♂ put Oil of Vitriol, then pour common ▽ upon it, and the filings will dissolve; filter the dissolution, and evaporate the Liquor *usque ad pelliculam*; set it in a cold place, and it will shoot into Crystals, which require no further Purification.

The Oil of Vitriol for this work is made thus: Take Danzick Vitriol, dissolve it once in ▽, filter and congeal it; then calcine it gently, until it becomes white. Then distill it in retorts S. A. forcing it very strongly at last. Dephlegm this Oil in a low cucurbite, and that which remains in the Cucurb. (which will be of a dark red) must be passed through a filter of wool in a glass funnel, and the wool will imbibe the unctuosity of the Oil, which if it were not separated from it, might hinder its operation.

To fix ☽ into ☉.

The 15th of November, 1660, Monsieur John Commandair told me, that Signor Lucca (from whom he now came) had taught him a shorter and easier way of doing his work, thus:

Take the Mother-liquor of Salt-petre, (which is the salt ▽ that remains after as much is shot into Nitre as will shoot) and let it run once through a filter of washed sand to purify it; then evaporate it to dryness: Grind the remaining salt very fine, and set it in a cellar, or other moist place to dissolve into ▽ by the Air; filter that by a woolen

Languette, coagulate, grind, dissolve, and filter
it. Repeat this, seven or eight times, that all
foulness may be severed from this fixed salt of
Salt-petre. Then it will easily give its pure
Spirits, and not before. Put this into retorts, not
above half a pound into each retort; distill first
with very gentle △ increasing it by degrees, at
last, strong △, as when you distill A. F. The
distillation will be performed in twenty four hours:
Then dephlegm it carefully; when the drops come
Acid, cease. In the meantime purify the fixed salt
remaining after the distillation, by grinding it
small, dissolving in humido, filtering, and
congealing. Repeat this twice or thrice; then put
one part of this fixed salt to three parts of the
spirit, and to this composition put a tenth part of
pure ☉, and though it were in an Ingot, it will
dissolve it speedily. Put this into an Egg, and seal
it Hermetically, and digest it, it will putrefy, and
grow entirely black; then pass all the due colours,
during which time increase the heat by degrees, and
when it requires strong heat, use Coal.

An Observation about Volatilised ☽

Monsieur de L'Oberie, and Mons. de la Nouë wrought the first process upon ☽ (which is after those upon ☉) that is in the handgrif of Basil Valentine, which makes the fourteenth Book of his Test[5]. But instead of a due Calx of ☽, they took one made with A. F. (the ordinary made of Vitriol and Nitre) and precipitated it with salted ▽ (Salt dissolved in Common ▽) and for the rest, did as the process teaches; which was reported to me thus. Put upon this Calx of ☽ (they had ℥iv.) (after it is well dulcified by often ablutions with fair ▽ till no saltiness or spirits appear to remain) so much fresh A. F. as to swim four fingers breadth over the Calx of ☽: Distill off the A. F. then cohobate again; do thus four times. At the last distillation give strong △, you will have a gray substance like marcasite. Beat it to powder, and put distilled vinegar upon it, to swim four fingers over it; digest two days, then boil it three or four hours, after which, distill away all the distilled Vin. and there should have remained blue Crystals, but they were white without tincture: So having failed in their expectation, they would reduce their ☽ back

---

[5] "Test" is the word used in the original 1682 text. -pnw

in a body, therefore dulcified it well with distilled vinegar and fair ∇, and put it into a Cruc. to melt with a little Borax, and a little Nitre, and a thick smoke flew away, and in the end there remained but ℥ij. of ☽.

Consider, if this course, and, if need be, digesting longer (at last) with distilled vinegar and Oil of Tartar, ✳, and salt of Urine, &c. Then distilling with Tartar and Calx-vive, might not make ☿ of ☽.

A Process From Monsieur Vignault,

With ☉ and ☿ &c.

Take ℥j. of ☉, àààte it with ℥iv. of ☿ grind this ààà, and wash it well: Then put it into an Earthen Pot with its cover to shut it very close, which cover must be like a funnel at the top: Put it in a gentle △ in sand for twenty four hours, then give it a strong △ for twenty four hours more, that the matter may ascend and descend; then take out your matter (loosening it from the bottom where it sticks fast) and grind it, and àààte it again with the same ☿, and proceed all as before. Repeat this work six times, always with the same ☿, which, by degrees will become earth, and will stick no more to the bottom; you must leave it in sand every time, twenty four hours before you grind it again; after the sixth time give it strong △, so that it may be red-hot in the sand for forty eight hours, and it will be a red powder, which multiply by mixing with it its weight of ☿, grinding and digesting it as before; and in three times twenty four hours it will be in powder; and if you will multiply it again, proceed as before, with equal weight of ☿. And to make it into a Tree, do thus: When you have made the àààà, and ground and washed it, then put it into a Matrass, which stop only with paper; then digest it continually, and the said ☿ will ascend and descend. When you see that at last it becomes hard and heavy, sticking to the neck of the Matrass, put it down with a quill, and it will become a Tree, which will be red. Note, That your ☿ must be well purified first, and then sublimed with ☉ and ☽, taking ℥ij. of ☉ to one pound of ☿, for it will be much the better, and will be sooner done. If you mix ℥β of ☉ with ℥β of the said powder, and grind it well with ℥ij. of ☿ revived from Cinaber, and animated with ☉, as is said, and digest it forty eight hours, you will do more in fifteen days, than otherwise in two months, and the ☉ will not stick to the bottom of the pot. You must continue the digestion as is said above, and at the end strong △. The ☉ will animate the ☿, and melt it, and reduce it into a Calx, for to *àààte* it with animated ☿, taking ℥j. of ☉ to four of ☿.

Fixation of ☽,

wrought by Father Bening de Baune, and by him communicated to me.

First, he animated common ☿ for this work, thus:

Take ℥iv. of Common Sulphur, melt it in an Earthen Poringer, then cast into it by little and little lbj. of ☿ (purified with salt and vinegar, and squeezed through Chambo-leather) stir it continually; then take it from the fire, and keep it stirring until it is reduced to a black powder, which grind, and add to it lbj. of ♂ in powder, and lb ℔. of Quick-lime also in powder; mix all together, and put it into a coated retort, of such a bigness, that a third part may remain empty. Distill it, and let the nose of the retort lie in a Poringer full of ▽ distill by degrees of △, as you do A. F.

55

the ☿ will distill into the ▽: Mix this again with new materials, and distill as before. Repeat this operation with the said ☿ seven times, every time with new materials.

Take of this ♀ ℥ iv. *àààte* it with ℥ j. of ☉; wash the *ààà* often, that the ▽ come from it clear, then dry it. Put this *ààà* into a Matrass, and digest twenty four hours in Ashes: Then take it out, and grind it in a glass mortar, and add to it ℥ xx. of the said ☿; grind them well together, then wash it and dry it, and put it into a retort, and distil over all the ☿ in sand.

Take ℥viij. of this ♀, *àààte* it with ℥j. of a light spongy-Calx of ☉; wash this *ààà* well with warm ▽ then dry it, and put it into a Matrass; seal it Hermetically, and digest it in sand the space of twenty four hours: Then grind it again with ℥viij. more of ☿, and digest as before. Repeat this operation once more with ℥viij. more of ☿, so that there be ℥xxiv. of ☿ to one of ☉. Put them into three separate matrasses, which seal hermetically, and put them to a suppressing heat in an Athanor, for the space of two months. Then put all into a retort, and distill it in sand, with a heat of suppression, so that the △ above be stronger than that below, and if any of the ☉ remain in the bottom of the retort, you must *àààte* it with twenty four parts of ☿, and distill it as before, until all the ☉ be distilled over. Repeat the same as before, until the ☉ has taken in sixty parts of ☿, and if it takes but twenty four of ☿, the ☉ will be better, and your ☿ will be animated.

Take ℥j. of Calx of ☽, and three or four of your ☿ animated, *àààte* them together, wash the said *ààà* with warm ▽, then divide it into two parts, and put them into two matrasses; seal them Hermet. and digest in an Athanor with very gentle heat for forty or fifty days, then increase the heat for forty or fifty days more: Then continue the digestion with the third degree of heat (stronger yet) unto the end of eight months, counting the time of the first and second degree already past. Then digest a month longer by the four degrees of △, which will make it nine months in all.

The Calx of ☽ is made of equal parts of ☽ and Regulus of ♂ melted together, and reduced to powder. Note, that the Reg. is not to be reckoned; so that you must take ℥ij. of this powder.

Observations.

The Athanor was of a digestive furnace, with a tower for the coals, and between both, there were two registers of heat, the one gave the heat under the vessels, and the other above: The Matrass stood in sand in a bason of copper, which held ten or twelve matr. At the beginning the △ was given only below, and so gentle, that the ☿ never sublimed. The bason with the matr. was covered with a cover like a Dome, and after that the heat was given also above, and that stronger than before: And it ought to be always continued without interruption. After nine M. digestion, all the ☽ will be transmuted into ☉, and besides that, you shall have an augmentation of a third part of ☉.

Note, that you must not put above ℥ij. of matter into each Matrass.

The ☉ which he used in this operation, was three times purified by ♁.

He told me since, that the greater proportion of Reg. you put to the ☽, the better your work will succeed, and you shall have the more ☉, and the sooner.

(Hartman) The said Father B. de B. was the Apothecary in the Convent of the Capuchins at Lyons:

He was an able Chymist, and had been for some years operator with the Chancellor of France, in his Laboratory. When I went from Paris to Italy, after Sir Kenelm's death, passing through Lyons, I went to see him in the Convent of the Capuchins, where I had some discourse with him concerning this work; he confirmed it to me, assuring me that he had done it, and that it was a real truth, and that is all I know of it.

A ▽ which changes ☿ as red as Blood,

which abides the Fire.

Make an A. F. of equal parts of Vitriol and Nitre, which cohobate and distill three times upon its Caput Mortuum.

Take of this A. F. ℥ iij. ℨj. of ☿, and ℥ ij. of Sulphur-vive; put all into a retort, let it stand twelve hours, then distill it, and cohobate so often, till you see the ☿ as red as Blood, which will be in five or six times; then bring it into a powder, which imbibe with Oil of Roman Vitriol, dry and imbibe it three times: Then divide this powder into eight parts; then take ℨj. of ♄, which put to Coppel, when it boils, put into it a Ducat of ☉, then put into it one of the eight parts. Drive it off, and you shall have ℥j. of fine ☉.

(Hartman) This process was written in the French Tongue; at the bottom was written Probatum, the 2nd of July, 1658. The process said it must be done on Thursday and Friday, and at the Full of the ☽.

Saunier's Work, as I wrought it.

1. Purify ☉ three times by ♄; then reduce it into a subtle Calx, by calcining it five times with Sulphur and ☿: Then burn S. V. upon this Calx, and reverberate it again, that all the extraneous Spirits may be driven away.

2. Sublime ☿ seven times with Vitriol and Salt, reviving it with filings of ♂ after every ☰ mation.

3. Make an A. R. S. A. out of the fixed salt, after the extraction of salt-petre, which after some days must be dephlegmed with great care, and rectified, so that it has neither phlegm nor terrestrial feces.

Dissolve ℥j. of your ☉ in as small a quantity of this A.R. as you can, keeping the vessel well sealed (and therefore it ought to be large) in a very gentle heat in B. M. where it must be digested

60

(after the dissolution) for some days: The dissolution being very clear, decant it from the white residue.

Dissolve 3ß. of fusible salt in as small a quantity of the said A. R. as you can (which is not done suddenly, but by digestion) and being clear, mix these two dissolutions together, namely, that of the ☉, and of the salt, and if anything ℞tate to the bottom; keep it in digestion with a gentle heat (the vessel close stopped) until all is dissolved and clear; then keep it in the same digestion for fifteen days. Then with a very gentle heat abstract the phlegm, until a spirit ascend; then cease, and put into the vessel 3ß. of the ♎mate before mentioned (in very subtle powder), shut the vessel again immediately, and put it in digestion as before, until the ♎mate is well dissolved. Then dephlegm again the dissolution; in doing of which you ought to attend very diligently, lest there come over some part of the ☉ and ☿, which now easily will be raised with the A. R. And this you may know, not only by the drops falling yellow, but also by trying with a white woolen cloth, which the drops will stain yellow if the ☉ ascends. Then seal it hermetically, and digest in horse-dung: After six months we opened the vessel, and with a gentle heat distilled off the liquor, and the remaining Golden

Salt we projected upon restricted ☽, and for ℥j. of ☉ we had seven. Another vessel, after twelve months digestion, rendered ℥x. of ☉ for one put in: And so to two and twenty for one.

I do not remember all the time precisely, but I should think, it would be better, after sufficient digestion in horse-dung, to coagulate the matter in dry heat until all be completely fixed, and then multiply the matter by the same process, as you did with ☉.

The fusible Salt is made thus: Dissolve salt (first well purified) in the said A. R.; distill and cohobate until it is fusible.

The restriction of ☽ you will find in a book published by John Saunier which he calls, the almost fixation of ☽, because it has the weight and sound of ☉.

(Hartman) This process was wrought by Sir K. D. himself, as the Title shows; it was written in Latin in his own hand, and the words are his own.

Abbot Boucaud told me at Paris, that he knew Sir K. had wrought it.

The Dane's Work.

Calcine plates of ♂ and ♀ with Sulphur; then grind them to subtle powder, which boil in ▽, filter and evaporate, *usque ad pelliculum*, and put it to Crystallize in a cold place. Then purify these Crystals by dissolving them in ▽, filtering and evaporating.

Make also a Sulphur of the said Metals, by boiling plates with Vitriol and ▽ in a kettle, and the Sulph. will adhere to the Plates.

Purify ☿ first by distillation and then by boiling it in an Earthen pot with Vitriol, ashes, and powdered glass well mixed together, and boiled until you see the ☿ appear upon the surface of the matter: Then let it cool, and grind all well together again, and boil it as before. Repeat this three times: Then take of this ☿ four parts, of the Sulph. of ♂ and ♀ *ana* one part, grind them well together until they be well incorporated; then ♎m and grind again what is ♎med with that which remained in the bottom, and ♎m as before. Repeat this seven times: Then is the ☿ prepared for this work.

Distill an Oil s.a. out of the Vitriol of ♂ and ♀ joined together, which will be Blood-red.

Make a light and spongy Calx of ☉, by calcining it four or five times with Sulphur and ☿. Take of this Calx ℥j. and of the ♀ prepared ℥iv. Make an *àáà*, which grind very well; then add of the Sulphur of ♂ and ♀ *ana* ℥ß, grind them well together with the *àáà*, then put it into a Matrass of such bigness, that three fourth parts may remain empty, stop it tightly with paper, that some moisture from the mercury may exhale, (which otherwise might hinder the precipitation of the mercury) give fire by degrees, first in ashes, and then in sand, but so gentle, that the mercury may never rise, but that it may be always in a disposition to sublimation, which you may know by a subtle cloud upon the sides of the glass, such as appears when one breathes upon a looking glass.

The end of the digestion is, when you see the matter converted into a very solid ☍tate and glittering, which endures a very strong △. Then take it out (being cold) and grind it with four parts more of the said ☿, and the same quantity of the said Sulphur as before; digest as before, until all be converted into a red ☍tate as before, except that it will be of a darker colour: Grind

64

this ♃tate with the Oil of Vitriol before—
mentioned until it be like a pap: Then put it into a
curcurbite, and digest for fifteen days, then
distill it, and the Phlegm will come over, and the
matter remaining dry in the bottom, you must grind
again with the new Oil, and proceed in all as
before. Repeat this so often, till the Oil comes off
as sharp as it was put in which is a sign of
saturation: Then digest this matter in sand until
all is resolved into a very red Oil in appearance
(which in a cold place will congeal into a hard and
brittle matter.) At last give a very strong △ for
three days, in which time the matter will be
entirely fixed, except a small quantity, which will
be exhaled.

Project this matter upon ☽ in fusion equal
parts. Thus far reaches my experience; but the Dane
told me, that if this matter were amalgamated again
with new ☿ prepared, and In all things proceeded as
before, taking this matter for the foundation,
instead of the ☉ which you took at first, it would
become a medicine, which in projection would convert
a great quantity of ☽ into ☉. And the oftner you
should do this, the more power it would have in
projection.

Out of ℥x. of this matter, and as much ☽, I
had ℥xvij.ß. of perfect ☉.

(Hartman) Dr. Astell, an English Physician, showed me a copy of this process, which Sir. K. D. had given him, who had assured him that he himself had wrought it, and that it was true: And having ℥x. of fixed matter, he divided it into ten parcels, and having melted ℥x. of ☽, he projected the said parcels one after another upon the same; then left it in fusion for three hours, then cast it in Ingot, which having weighed, he found the quantity of ☉ above mentioned.

Opus Magnum ex Virginea Terra.

Take reddish rich Virgin Earth in ♈, impregnate it with ☉, ☽, serene and dew, till the end of May: Then imbibe sprinkling with dew gathered in May, and dry in ☉, expose all night to the ☽ and Air, securing it from Rain. Still when it is dry, imbibe and turn the earth often. Continue this till ⚌mation. The hot ☉ (especially in the Dog-days) will make a pure salt shoot up, which mingle back into the earth, by turning it all over. Then distill by graduated △ as A. F. forcing all the spirits over at last; you must give forty hours △, extreme at last. Put all the liquor and salt that comes over, to digest and circulate a month *in fimo*, in a great balloon close shut. Then separate the several substances out of this Chaos; first, comes an extreme subtle, ardent, AEthereal spirit, then white ones with veins like S. V. then Flegm. Thus far in B. in a cucurbite, then in a retort: Then white fumes, then red ones, and a reddish brown salt remains in the bottom, and a Volatile salt will be ⎯med about the neck of the retort, as also to the head and sides of the cucurbite. Then purify every substance by itself; the fixed salt by solutions in the Flegm, filtrations, and congelations, till it be perfect pure, clear, and cast no more feces: The volatile salt by often sublimations: The first spirit by thrice distilling, and the fixed white and red spirit likewise, both together. Now join all the three parts, beginning with the fixed salt, whereof take three parts, and one of the fixed spirit; digest eight days, distill in ashes, and the liquor will come off like Flegm. Imbibe with more fixed spirit, and repeat this till all of it is coagulated with the salt. Then put one part of this to three of the ✳, taking it all, and humect them with the volatile spirit. Digest eight days or longer, then distill in a cucurbite; a stinking flegm will rise, and a pure salt ⎯m up, and if any spirit distill over, keep it, putting it to the rest of the spirit. Then add more of the fixed salt to it which has not ⎯med, making it one third to the ✳ which humect with spirit as before, circulate and ⎯m, and the

67

✳ will be increased. Do thus till all the fixed salt be ⚏med. Circulate the remaining volatile spirit with the ✳, till all the spirit be converted into ✳, and nothing but a stinking flegm comes away. Then ⚏m this salt by itself, till it leaves no feces, and is most white, transparent, and pure, which will be in four or five times.

Take seven parts of this ✳, and one of pure ☉ in leaf, seal it hermetically, and digest in B. The matter will become a green ▽, like an emerald, with an Oriental esclat: (and in a retort will pass all over, leaving a few grains of brownish-gray stiptick Earth, like tobacco-pipe Earth). And after a while black like ink, and continue so two and forty days; when the blackness begins to wane, put it in dry △ in an Athanor. It will pass the colours, and become a red Elixir, and is now best for health; but it will not have good ingression into metals, till it has been multiplied four or five times with new ✳, taking every time after the first, only three to one; and it will be done every time after the first in a shorter space. After every fixation of the multiplication, and the first also, give strong △ for three days, and a black earth will separate from the red powder, lying like a cake under it. Before you project upon inferior metals, ferment anew with three parts of ☉ to one of the Elixir, giving three hours of extreme fusion, and all will be red powder.

You may proceed in the same manner for ☽.

Note also, that when the work of ☉ is at the white, it will project upon inferior metals, to make them like ☽, but in truth white ☉, enduring all the trials of ☉.

If you digest in B. V. ten parts of Pearl in Powder, with one of the perfect ✳, it will become an Oriental Liquor, whereof a few drops is admirable for health.

If you take four parts of such ✳, and grind it well with pure red Coral in powder one part, and ♎m, putting what rises upon as much of fresh Coral, repeating this four or five times, the ✳ will be red like a Ruby, and an admirable medicine. AU the Corals will dissolve in a Cellar.

If you grind one part of it with ten parts of green Venice Talc, and put distilled dew upon it, six fingers over, and digest *in fimo*, all the Talc will dissolve, and a splendid Oil of rare effects swim upon it.

(Hartman) Sir Kenelm D. said, that a person of quality beyond the Sea (whom he named) wrought this process, and it happened at that time that his wife was dangerously sick, and like to die; she was given over as a dead woman by the ablest physicians: Upon

69

that he opened the vessel, and gave her one grain of the Elixir; whereupon she recovered, and lived many years after it in perfect health.

This process, and Saunier's[6] work were together in a small bundle of papers tied up together by itself; upon the outside of it were written the following words, perfumes, curiosities, my great Arcane of this Note.

A Minera of ☉

wrought by a Person of Quality in Champagne.

Take Sulphur-vive lb ℔. melt it in an Earthen Poringer, then squeeze into it lbj. of ☿; stir it continually until the ☿ appears no more in the Sulphur. Then let it cool, and grind it to powder, which digest in a Matrass for two days with a strong △. Then take it out, and grind it again; add to it its double weight of filings of ♂; mix them well together, and put them in a retort, and distill over all the ☿: Mix this ☿ again with new Sulphur melted as before; digest in a Matrass as before for two days, in the meantime grind the filings of ♂ (that you distilled the ☿ from) and wash them well from

_____

[6] Is this the French priest who discovered a great secret? HWN

all the foulness and blackness: Then dry them and grind them again with the Sulphur and ☿, and distill them in a retort as before. Repeat this so often, till the filings of ♂ come to be of a yellow Golden colour, which will happen at the seventh distillation: Then take this ☿ and put it in a retort, and distill only ℥j. of it, and with the remaining ℥viij. Make an *ááá* with ℥j. of ☉, digest this *ááá* in an Athanor for nine months, it will pass all the due colours, and will become a Miniera as follows. To this ℥ix. of matter put ℥iij. of ☿ prepared as before, and digest, and in six weeks you shall have ℥xij. ready to melt: And to these ℥xij. add ℥iv. more of ☿, and digest, and in six weeks you shall have ℥xvj. of Miniera. Note, that you must always use a ☿ prepared, as was said for the multiplication of the Miniera: For if you should take crude and unprepared ☿, you would have but an ordinary [illegible]tate after one or two multiplications.

Note, That the filings of ♂ is to be changed after three times, and new to be taken, which is to serve also three times: After which six times, you must join both the parcels of filings, and use them both at the seventh time, and if the sign given you (of the yellow Golden colour) happens not at the Seventh time, continue and repeat your operation, with all your filings, until it does appear. When your Miniera is completed, it will be a deep-red powder, very shining, and at every time it is to become such: If you multiply it with crude ☿, it will lose its lustre after twice, and not increase in fixed metal.

The first time, you must put into one glass no more than ℥j. of ☉, and ℥viij. of ☿: But when the Miniera is made, you may work even to fifty ounces in one glass, keeping always your due proportion.

Fixation of ♄ into ☽, with good Profit.

Melt lbj. of ♄, then put in ℥ß of ☽, and some scories of ♂ and a little red Arsenic; keep it in a strong △ for three or four hours or more. Then the crucible being cold, break it, and take out the matter, and put it in a new cruc. which must have a little hole in the bottom; put this cruc. in a wind furnace, and melt the matter again, putting under the furnace a bason with ▽ to receive the matter as it melts and runs through the cruc. Take this matter and melt it again with the same quantity of ☽, and new scories of ♂; keep it in fusion as before. Reiterate this operation ten or twelve times, until the ♄ is very hard, being impregnated with ☽; then put it to Coppel with ℥j. of ☽ to every lb. of this mixture.

The goodness of the operation consists in the fixation of the ☿ which is in the ♄ by the Sulphur of ♂: Therefore you must keep the matter a long time in fusion, that the Sulphur of ♂ may act strongly upon the said ☿.

To fix ♀ of ♂, or the Common ☿.

Take ℥j. of ☉ in leaf, and ℥iv. or v. of ☿: Make an *ááá*, which put in a retort, and digest it in horse-dung for eight days, then distill in sand, giving strong △ at last, and the ☉ will go over with the ☿, and if any of it remains in the bottom, *ááte* it with the same ☿, and digest three or four days, and then distil as before, and all the ☉ will go over with the ☿, and you shall have a ♀ well animated.

Take ℥iij. of this ☿, *ááte* it with ℥j. of ☉; grind the *ááá*, and put it in a Matrass half luted; digest for eleven days by graduated △, and all will be a red powder.

Take ℥iij. of this powder, and project it upon ℥j. of ☉ in fusion, and all will be transmuted into ☉.

Then take the remaining ℥ of powder, and *ááate* it with ℥iij. of the ☿ animated; digest as before, and in nine days your powder will be perfected as before. Take these ℥iv. of powder, and unite it with ℥xij. of new ☿ animated; digest without ☉, and you will have a perpetual Miniera; part whereof you may reduce to a body when you please, by projecting it upon ☉; and the other will serve for a ferment, which will never fail, being itself all ☉.

This ☿ animated may be fixed without ☉, by a gentle heat, being itself a liquid ☉; but to shorten the work, you may add ☉.

A Reality upon ☽.

Take ℥ij. of ♀ in thin plates, and ℥j. of small nails, put them in a cruc. in a furnace, and when they are very red, cast in some Sulphur upon them at several times, that they may melt well; when they are like paste, cast in some ♄, and stir it with an Iron Rod to make them well incorporate: Leave it in good fusion for five or six hours, stirring it sometimes. Then take out the cruc. and let it cool;

then break it, and you shall find but a little Reg. at the bottom, but many yellow lumps at the top, which beat to powder. Then melt ℥ij. of f1ne ☽, and project ℥iij. of the powder; stir it with an Iron Rod, keep it in fusion for eight or ten hours. Then put it to Coppel, and separating ▽[7], and you shall have fine ☉.

(Hartman) This process is also confirmed with a Probatum.

Fixation of the ☿ of ♂,

as Monsieur de la Noue wrought it in Paris.

Take ☿ of ♂ and ☉ *ana* ℥j. Oil of Vitriol ℥vj. Distill to dryness; take what is ♎med, and join it again to the feces, and put the Oil upon it again that distilled over; distil as before. Repeat this so often, till nothing more ♎m, distilling every time in a new retort; at the twelfth or fifteenth distillation, all the matter will remain in a red powder.

Take Sulphur-vive, and ashes of Alicant *ana* equal parts, of which make a lixive with common ▽;

---

[7] Both the original 1682 printed text and the R.A.M.S. version used ▽ (water) here, in error. I have inserted the correct symbol: △ (fire). -pnw

filter and evaporate, and you shall have a Sulphurious Salt: Take of this Salt and of the said powder *ana* gr. vj. ☿ of ♂ ℨj. filings of ☉ ℨij. mix and grind all well together, and put them into a matr. with a long neck; make a △ about the middle of the neck of the matr. in an iron pan with a hole in the middle through which the neck of the matr. may pass; let this △ be stronger than that below; continue the △ for six hours: Then cast your fixed matter into a Bath of ☉.

Preparation of the Powder,

with which Claudius de Montrouge, and Abbot Oberye at Paris fixed ☿ of ♂.

They melted ℨiv. of Sulphur in an Earthen Poringer, then they squeezed into it through a leather ℨj. of ☿ of ♂ made of Regulus of ♂, ⚹, and ☿ ♎mate (the ☿ of ♂ without addition had been better, but they had none) and whilst the one squeezed the ☿ into the Sulphur, the other kept stirring continually with an Iron Spatula so long until the ☿ did no more appear in the said Sulphur, and that all was converted into a grayish Citrine Powder (the colour is variable, according as you

govern the △ sometimes it will be red like
Cinaber.)

   To this powder they took ℥j. of ☉ in Calx, and
℥j. of the Salt that is found in the pots at the
glass-houses, which salt they dissolved, filtered,
and congealed: They ground all well together, the
powder, the ☉, and this salt: Then they put all
into a retort, and put upon it ℥xxiv. of good Oil of
Vitriol well rectified; to this retort (being put in
sand) they adapted a large glass receiver, the
junctures being well luted, and the lute dry, they
distilled by degrees of heat, at last gave strong
△. It was ten or twelve hours before the Oil came
over. All being cold, they broke the retort, and
took out the matter which remained in the bottom,
which they did put into a new retort, pouring upon
it the Liquor with the flowers of Sulphur which were
in the Recipient: Then joining again the receiver,
and luting well, and the lute being dry, they
distilled as before. They reiterated this operation
twenty times, grinding every time the matter, and
joining it with the Liquor and Flowers.

   At the twentieth distillation, the small
quantity of Liquor that came over, was almost all
flegm; then they took out the matter that remained
in the retort, and put it into a Vial, which they
stopped very close, and kept it in a dry place,

77

because that so soon as it felt the Air, it grew moist.

With this powder they fixed the ☿ of ♂, which being mixed with the Calx of ☉, and held in the hand, grew so hot, that they were not able to hold it in their hands, no more than a piece of Iron red-hot, as every one of them made experience, casting it into a bason full of ▽, which they had standing by for that purpose.

They wrought the said fixation in an iron barrel of a gun, thus. They put about sixty grains of the aforesaid ☿ only (because they had no more) into the said barrel, then they gave the △, first above for two hours, and afterwards below for one hour, keeping that above always stronger than that below; then they heard the said ☿ begin to roar, and make a noise in the barrel; then they cast into it a little more than one grain of the fixative powder, wrapt up in paper; and then they continued the △ for seven or eight hours, after which time they heard no more noise at all; then they judged that the work was done, and let the △ go out; and the barrel being cold, they found about twenty grains of good ☉, which endured all the trials of ☉.

(Hartman) This relation is of Sir K. himself, written in the French tongue.

A Process to fix the Common ☿ by the Salt of ♄;

wrought by Captain Ziegler at Ments,
and sent me by him.

Melt it in an iron pan, let it be red-hot, then cast in some salt, stir it until it is reduced to powder; sift this powder finely, and that which will not go through the size, must be calcined as before: Then edulcorate this powder with warm ▽, and you shall have a Calx as white as Ceruse, which put into a Matrass, and extract the salt out of it with distilled vinegar s. a. after three or four days digestion, decant the distilled vinegar, and put on fresh; digest as before, shaking the vessel often: Repeat this three or four times, or so often, till the Sp. of V. has extracted all the salt. Then put all your Sp. of V. together and filter it, then distill it off in a retort, until you see the salt of ♄ remains in the bottom like deep-red Oil, which being cold, will be white like sugar-candy: Grind this salt, and put it into a Matrass, and extract it with Sp. of V. as before. Repeat this purification three or four times, and you shall have a salt of ♄ well prepared for this work.

An A. F. to be used in this Work.

Take salt ℥iv. Nitre lbj. mix them well together with lbij ß. of powder of bricks; put all in a retort, and distill by graduated △, forcing over the spirit strongly at last. The distillation will be performed in sixteen or eighteen hours.

Take ☿ seven parts, fine ☽ one part; make an *ààà*, which put into a retort, and pour upon it so much of the A. F. as may cover it a large fingers breadth: Let it stand twenty four hours, then distill it in sand; when it is cold, cohobate the distilled A. F. upon it again, and distill as before. Repeat this three or four times; then break the retort, being cold, and take out the *ààà*, which grind to a fine powder, and put it in an iron pan, and hold it over a coal △, stirring it continually with an iron rod, until it is almost red-hot, and that it be converted into a red powder, like red ⊽tate. Take of this red powder two parts, and of the aforesaid salt of ♄ one part, reduce them to a fine powder, which put into a Matrass, and digest it in sand for eight days: Then put it to Coppel, and you shall have half your *ààà* fixed into fine ☽.

(Hartman) When Sir K.D. was at Franckfort in Germany, where he lived a year and half, in the year

1659, he went often from Franckfort to Ments (being four German Leagues distance) to visit the Prince Elector there: Then he conversed also with this Captain Ziegler, who was a famous Chymist. And when Sir K. returned to England about the time of the Kings happy Restauration, the said Captain sent him this process written in the German tongue, assuring him that he had done it: He said, that the ☽ which he got, he put to separating ▽, and he had some ☉ out of it. He said also, that he thought this salt of ♄ would fix ☿, in ☉ if the *àà* were made with ☉ instead of ☽.

A Work upon Cinaber,

wrought by Monsieur Sauvage.

Take Nitre and ✳, *ana*, which dissolve in rain ▽; filter and evaporate to dryness: Then grind this double salt to subtle powder; take a large Crucible, in the bottom whereof put a bed of Quick-lime in powder, upon that put a bed of this salt, cover it with another bed of Quick-lime the same quantity as before, taking two parts of Quick-lime to one of salt. Cover the cruc. with another, without luting them; put this in a bakers oven after the bread is drawn, let it stand as long as there is any heat in

the oven; when the oven has been heated again, and
the bread drawn, set it in again; do this three
times: Then keep it in a strong $\triangle$ for six hours,
and being cold, take it out, and put it into ∇ and
let it boil in an Earthen pot eight or ten walms[8].
Then filter it hot, and evaporate to a dry salt,
which put in a strong bottle, and keep it close
stopped in a dry place. Then take two parts of this
salt, and of salt of ♄ one part, mix and dissolve
them in distilled vinegar.

Then take Cinaber, pulverize it, and make a
paste thereof with the yolk of an egg; of this paste
make little cakes in the shape of the heads of
horse-shoe-nails; make them pretty thick, and put
them in an Earthen pot, pouring upon them of the
aforesaid dissolution, so much as may cover them the
breadth of three or four fingers; boil this together
until it come to be like Honey. Put more distilled
vinegar upon the cakes, and boil it as before.
Continue this for three days; then wash the cakes in
fair ∇, and you will find them something
metallized. Filter the ∇, and evaporate to a salt,
which will serve again for the same use, adding salt
of ♄.

---

[8] A walm is a surge upwards of boiling water, when a circular
wave of water rises from the bottom of the pot and breaks the
surface in a surge. -pnw

Take of the fixed salt without salt of ♄, and
of good Venice Ceruse *ana* equal parts, grind and mix
them well together; then put a bed thereof about the
thickness of a Crown, into an iron box, then put a
bed of plates of ☽ upon that, and then the powder
again upon the ☽, the same quantity as before; upon
that put a bed of your lumps of Cinaber, then
powder, then plates of ☽, then the powder again;
and thus continue stratifying until your box is
full, the powder being the first and last. Then put
on the cover of the box, which you must fasten, and
secure it well with iron hooks. Then you must have
another box of iron, made big enough to contain the
first, and that there be the space of a fingers
breadth between, at the bottom, on the sides, and at
the top; the boxes must be square, and you must have
two iron hoops made in the shape of a Crown with
crankles; put one of them into the bigger box,
turning the teeth or crankles downwards, upon which
set the lesser box; put some pieces of iron on the
sides, to keep the lesser box at an equal distance
from the sides of the bigger: Then put on the other
hoop upon the lesser box, keep it down with some
heavy thing whilst you pour in some melted ♄ into
the bigger box, so much as may cover the lesser box
a fingers breadth. Then put on the cover of the
bigger box, and fasten it with iron hoops and wedges

to keep it closed. Then the box being yet hot, put it into an Athanor where the $\triangle$ is kindled, let the registers be shut, so that there be but a very moderate heat, such as where you may endure your hand; continue the first degree for three days, so that all that while the ♄ may be but melted, then increase the heat for three days more; and so increasing the heat every third day, continue in all three weeks; the last three days the $\triangle$ must be very vehement. Then let all cool, and take out your lumps, and reverberate them with very gentle heat for twelve hours, and they will be of a whitish-gray colour. Then melt ♄ in a cruc. and cast these lumps into it, digest this matter together for three days, then put it to Coppel. Note, that if you cast this mass into melted ☽, and digest it three days before you Coppel it, you shall have more profit than if you test it without digesting it.

Note also, that if you will continue your work, you need not use any more plates of ☽, but only the cakes as they are, and before they are reverberated, using them instead of the ☽, being pulverized, and they will be the more fixed, and the profit will prove very considerable.

You must have of ☽ and Cinaber *ana* ℥vj. and of the double salt and Ceruse *ana* ℥iv.

Tincture of Mars.

Dissolve filings of ♂ in A. F. made of
Vitriol, Nitre, Allom, and Cinaber; then pour upon
this dissolution distilled vinegar, double the
weight of the A. F. shake it well together, and
digest in B. for three days, then decant the clear,
and filter it; evaporate it gently. Then grind it
with two parts of ☿ ♎mate. Sublime the ☿ from it
four times; then dissolve it again in distilled
vinegar, and evaporate it gently; then dissolve it
in distilled Rain ▽, and congeal it gently: Repeat
this last solution till it is not corrosive upon the
tongue; then in ℥iv. of Rectified Spirit of Vitriol
dissolve ℥j. of this Sulphur of ♂, and ℥ij. of
Sulphur of ☉ made the same way, except the first
solution of the ☉ which must be an A. R. made of
salt, Nitre, and Vitriol; mix these two last
solutions together, and digest *in fimo*, then
coagulate it gently, dissolve again in spirit of
Vitriol, and coagulate: Repeat this seven times, and
if any feces remain at last, leave them out. Try
this medicine upon a hot plate of ☽, if it
penetrates and tinges it thoroughly without smoking,
it is a sign of its perfection; but if it smokes,
you must dissolve it again, and gently coagulate.
Then melt ℥j. of ☉, and cast upon it by little and

little ℨj. of this medicine, and when all is entered and incorporated with the ☉, cast it in Ingot, and you shall have a matter as brittle as glass, and transparent like a dark Granade stone, and fusible as ♄. Then melt fine ☉ and fine ☽ *ana*, and project of this medicine upon it, and you shall have pure ☉.

To fix a quarter part of ☽ into ☉.

Take filings of ☽ ℨj. *áámate* it with ℨiv. of ☿, put this *àà* in a retort, and distill off the ☿; take the ☽ and *re-ááámate* it with the distilled ☿; distill as before. Repeat this three or four times, and the ☽ will be a powder impalpable. Take ✳ and Cinaber *ana* ℨjß. ♀⚖ mate ℨß. grind and mix them well together with the ☽. Then ⚖me it with gentle heat, mix what is ⚖med with that which remains in the bottom, and sublime as before. Then take both feces and ⚖mate and mix it with Sulphur of ♀ and Crocus Martis, and of a Regulus made of ♄, ♂, and ♀ *ana* ℨß. grind all together with a little ✳. Then sublime it four times with gentle heat, adding every time a little ✳, because it opens the body of ♂

and ♀, and unites them with the ☽. Then grind all well together, and digest it in the following ▽. Take Nitre, Vitriol, *ana* lbj. ♂, Sulphur, Verdigrease, and Auripigment *ana* ℥iv. Make an A. F. of this, s. a. Or take common A. F. lbj. distill and cohobate it three or four times upon the said materials, giving strong △ at last: Then put your powder into a retort, and pour upon it so much of the A. F. as may cover it the breadth of three fingers, distill it off with a gentle △, then cohobate and distill three or four times: Then put fair ▽ into the retort, and digest for five or six days in sand; then evaporate to dryness: Then take out this matter and pulverize it, and weigh it. Then melt as much ♄ as you have powder, and cast your powder upon it by parcels, melt it with a strong △, then let it stand in the △ until the △ go out of itself; then take it out, and you will find a Regulus in the cruc. which Coppel, and then put the ☽ to separating ▽ and you shall have a fourth part of fine ☉.

A Work with Butter of ♂

The work, which Monsieur Perdussin of Lyons communicated to P. A. Dieudoné, is to make a butter of ♂ with ♂ mineral, and ☿ ♎ mate, *ana* lbj. Of this take ℥ij. and digest it in a Matrass sealed hermetically in an Athanor, and it will putrefy, growing as black as pitch; then pass the colours. That done, take one part of leaf ☉, and three of this powder; grind them well together, and digest as before, it will become black as at first, and pass all the colours. This proportion of ☉ for ferment, you may divide into several parcels, for several times, so each revolution will be shorter, when the whole dose of ☉ has fermented the first stone: This product serves for ferment to multiply in quantity and quality. The P. wrought the first part, and had perfect putrefaction.

An Excellent fusible Salt.

P. Benin de Beaune made his fusible salt thus: Decrepitate and reverberate salt, then dissolve it in fair ▽ filter and congeal. Repeat all this work four or five times: Being perfectly pure, dissolve it in spirit of vinegar, filter and congeal; repeat this with distilled vinegar once again: Then it is perfectly fundant.

Another fusible Salt.

Dissolve salt in rain ▽, filter and congeal; when the ▽ is almost evaporated, and that the salt falls to the bottom, take it out with a wooden spoon by little and little, until the ▽ is exhaled: Grind this salt (being very dry) and reverberate it in an Earthen vessel close luted; let the vessel be red in the △, but the salt must not melt; so soon as you see the vessel red, let it stand until the △ has gone out. Then grind it and reverberate it as before; dissolve and congeal as before. Repeat this until it is perfectly fundant. Note, that you must not decrepitate your salt.

An Operation with a Martial Regulus of ♂

wrought by Monsieur Toysonnier.

He made a yellow Martial Regulus thus: Ignifie ℥iv. of Nails in a crucible, then put upon it ℥viij. of good ♂ , and give strong △ in a wind furnace, to make all melt well, which to promote, cast in

some saltpetre, then cast it In an ♂ Horn, and
separate the feces from the Regulus. Ignifie ℥ij. of
Nails more, and cast thereon the feces (this work
must be done presently after the first) adding salt-
petre to make all melt well and clear: Then cast it
in a Horn, and separate the scories from it, and
wash it clean; it will be first white, but after a
day or two will be yellow within as well as without.

Take of this Reg. and of ☽ *ana* ℥ß. melt them
well together (he poured a little ☿ in them when
they were near ready to congeal, and stirred with an
iron rod, but the mass took in little above ℥j. of
☿). Beat it to powder, add to it eight or ten parts
of ☿, and grind exceedingly till they incorporate,
(which required about twelve complete hours, often
heating the matter and instruments). Then squeeze
away so much ☿, that there remain only six parts;
digest it three days in sand by degrees, at last,
very hot. Put the remaining Calx to Coppel with four
charges of ♄, adding a little fresh ☽ to make it
work better. Put the mass, *au depart*, and you shall
have twenty six gr. of good ☉.

(Hartman) The said Monsieur Toysonnier was Sir
Kenelm's operator; he was a French-man, and a very

able Chymist. Sir K. brought him over with him from Paris, 1660.[9]

Butter of ☿ to Extract the Tincture of ☉.

Digest butter of ☿ six weeks or two months, and then put it upon a well-opened Calx of ☉ and digest it, and the B. will extract the tincture of ☉, which digest, etc.

To fix ☽ into ☉;

given me by an Intimate Friend, who told me that he wrought it as follows, taking his hints out of Lully's Experiments.

He made a Mercurial ▽ as he teaches, by his vessel with three bouls in three furnaces (which Mercurial ▽ will return again into running ☿ after a little digestion) and to this he put some pure white salt of Tartar, and some ☿⎺ᴠ⎺tate, that had been ⎺ᴠ⎺tated by itself with three or four months digestion, and some Calx of ☽ exceedingly well opened, and very subtle: This he digested a good while, and drew off the ▽, and cohobated several times, after which he did put some tincture of ♂

---
[9] "Our Luna" per Bacstrom. HWN

unto it, and digested and cohobated anew; and in the end he found almost all the ☽ converted into ☉, that endures all trials, but it was a little pale.

In Lully you may find directions to make all the things that were used in this work. The salt of Tartar, was but the fixed, reduced to its highest purity; but it should have been Volatilised, and made to pass over with the Mercurial ▽, to acuate and animate it. He believes the great work is to be made with a Mercurial ▽, animated with a volatile salt of Tartar, to serve for a Menstruum or Alcahest to dissolve ☉ and ☽. Weigh well what Lully said of these things.

Mallus his Process to fix ☽:

Wrought by Monsieur Ferrier, and given me by him, 1660.

Take an A. F. made of equal parts of Vitriol and Nitre, pour of it upon Sulphur and ✳, *ana* (four parts of A. F. to one of powder) distill it off to dryness, and make sublime what will. Melt ℥iv. of ☽, cast upon it ℥ß. of sublimed salt when the ☽ is in good fusion: After it is entered cast in Ingot, melt again, and project a new packet of salt, doing

92

all as before: Do this four times, so spending ℥ij. of salt upon ℥iv. of ☽. Then put it *au depart*.

To fix ☽ by a Mercurial Water.

Make Mercurial water by means of an earth retort that has a pipe or spout behind in the upper part, through which you cast in the ☿ when the retort is red-hot. Take of this ▽ (well rectified) ten parts, and one of a well rectified Oil of Vitriol; distill them together, till they are perfectly united: Then take of this Menstruum ten parts, and one of a well calcined ☉; digest them together in a Matrass (sealed hermetically) until the ☉ is well dissolved: Then take it out, and put the matter into a low cucurbite, and distill until the drops come Acid. Then let it cool, and put the matter into a Matrass, seal it hermetically, and digest in an Athanor, until it is perfectly fixed into a red powder.

Monsieur Bertault's Tincture of ☉ by Venus.

Take sulphur and borax, *ana*, melt them together three times, grinding the matter every time; then melt ☉ and ♀ *ana*, and cast upon them of the said composition, until the ♀ be reduced to *aes ustum* then cast in Ingot, and beat it with a hammer, to cause the *aes ustum* to scale off from the ☉. Then melt this ☉ again, and project as before. Repeat this three times, and you shall have a ☉ as red as blood, and this tincture will hold the test.

Note, that when you beat your ☉, if you see that the *aes ustum* does not all scale off from the ☉, you must melt it again, and project more of your composition until it comes all off in scales, and is all separated from the ☉, which it ought to do at every time.

To fix ☽ into ☉.

Distill from ☿⚖︎mate a spirit, wherein dissolve an *àáà* of ☉ and ☿ into a white powder, which digest in ashes until it be as red as Cinaber. Then dissolve it in A. R. into a red water, which reduce again to powder, which project upon ☽.

Another Tincture of ☽

Dissolve ℥j. of ☉ in A. R. and ℥iij. of ☽ in A. F. Precipitate them, and then unite them together, and distill their spiritual essence, which by degrees of heat fix into a blood-red powder, which tinges ☽ into ☉.

An Operation with ☉ and ♀ of ♂:
Wrought by Monsieur Chambulan, and given me by him.

Take lb iij. of good salt of Tartar, calcine it, so that it is glowing hot for twenty four hours, in a pot close luted; then dissolve it in flegm of brandy, filter, and evaporate; calcine it again as before, dissolve and congeal as before. Repeat all this work four or five times, or until it leaves no more feces in the filter. Then calcine this salt again for six hours, and then pulverize it whilst it is yet hot, and put it in a large cucurbite, and pour upon it by little and little of good *Nants* brandy, so much as may cover it the breadth of four fingers, cover it with a blind head, or with another cucurbite, that may enter into it, lute well the junctures, and digest in warm sand for six days; then put on a head with a Limbeck and a Recipient, and distill over with a gentle heat all the S. V. then let it cool, and pour on fresh brandy, digest, and distill as before. Repeat this operation five or six times, or so often till your salt of Tartar remains in the bottom like a red and transparent Oil, which will be very fiery and penetrating, reducing all metals into running ☿, being first duly prepared; keep this Oil close stopped.

Then take lb viij. of the ashes of burned vines, whereof make a strong Lixivium with lb xx. of fair ▽ then in lb xij. of this Lixivium; dissolve lbj. of salt of Tartar, filter this dissolution, and digest it in sand with a strong △ for some time; then cast into it lb j. of Regulus of ♂, that has been melted and purified six or seven times with Tartar and Salt-petre, and then reduced to a subtle powder: Make it boil for six hours, until the Lixivium is very red and stinking; then let it settle and cool, and decant the clear, and wash the powder with fair ▽, then dry it, and grind it upon a stone, imbibing it with the red Oil of Tartar before mentioned, until it is like a Pap, then dry it, and imbibe it again, and grind as before. Repeat this so often till the powder has taken in double its weight of the said Oil of Tartar: Then put this matter in a body with a blind head, lute well all the junctures, and digest *in fimo* for twenty days;

then take it out, and you will find your powder
converted into running ☿, which wash well with hot
▽, then with salt and vinegar, and then with fair
▽, then squeeze it through Chambo-leather. Then
take ℥x. of this ☿ of ♂, and ℥x. of common ☿ that
has been distilled over in a retort with Tartar and
Quicklime, and then washed with salt and vinegar;
mix these two Mercuries together and squeeze them
through a leather, then put them in a cucurbite,
lute another cucurbite upon it, and digest *in fimo*
for fifteen days, then put a head to it with a
Limbeck, and distill in ashes, and all the common ☿
will distill over drop by drop as ▽ and the ☿ of ♂
will remain in the bottom like a clear Oil, and will
be of a fragrant scent: Rectify the ▽ in ashes, and
the Oil with a stronger △ in sand, and keep them by
themselves. Then melt ℥ij. of ☉, and ℥j. of ☽, then
cast in Ingot, and beat it into leaf, or reduce it
into fine filings, and make an *áâá* with ☿ distill
this *áâá* in a retort until all the ☿ is distilled
over; then put this *áâá* in a Matrass, and pour upon
it ℥x. of the Mercurial Water before mentioned:
Digest it, and in a few hours all will be dissolved.
Put this dissolution in a retort, lute a Recipient
to it, and having luted well the junctures, distill
in sand, and all will distill over except a few

97

black feces; dephlegm it with a gentle heat in B. M. distilling until nothing more comes over. Take of that which remains in the bottom of the cucurbite ℥ iv. put it in a strong Matrass, and put to it ℥viij. of your Oil of ☿ of ♂ seal it well, and digest it with a Lamp △ in ashes, and in forty days all will be fixed into a red stone; then take out the Matrass, and put it to a strong △ in sand to Sublime it for twenty four hours, and all will melt like an Oil, which will congeal in a cold place into a red stone.

Fermentation.

Take ℥iv. of this red stone, pulverize it, and stratify it with ℥j. of ☉ in leaf between two crucibles well luted; put this to a Circulatory by degrees for six hours, then cover it with coals, so that it may melt and unite well together: Project ℥j. of this powder upon ℥x. of boiling ☿ (well purified) and all will be converted into a medicine, which will project upon a great quantity of ☿, transmuting it into fine ☉.

Elixir of ☿, ☉, and ☿.

Take good mineral ☿, mortify it with redicated vinegar; then separate its Quintessence with pure S. V. With that Quintessence dissolve ☿ *duplicatum* of ☿, that both become an Oil, which unite with a subtle Calx of ☉ and bring them to an incombustible Oil, which will transmute ☿ into ☉.

Elixir ex ☉ & ☽.

Dissolve ☉ (well purified by ☿) in A. R. then reduce it into a blood-red Oil with radicated vinegar, and Tartarised S. V. Then with this Oil imbibe a natural Sulphur of ☽, and fix them by graduated △. This is a high Projection upon ☽.

Elixir Album.

Sublime ☿ three times from Vitriol and Salt-petre, then in hot sand fix it so, that in strong heat it may not rise, which may be performed in three weeks' time: Then Calcine it in a close Reverberatory △, and it will be ready for solution. Then take the water which distilled over in subliming the ☿, and dissolve in it a little ✳, and

☿⚖mate; with this solution mix Calcined Vitriol to the thickness of Honey, digest *in fimo* one and twenty days: Then distill by degrees a little at a time (for it yields a very fiery Spirit) let the Recipient be large. When all is come over that will, rectifie it; then in this spirit dissolve the aforesaid fixed ☿, so is the Menstruum prepared.

Then take a white Calx of ♃, pour upon it so much of this Menstruum as will cover it, let it stand eight days as before; Repeat this till the Calx will take in no more of the said Menstruum, then let it stand till it becomes first black, and then white, Subliming itself above the Caput Mortuum, from which carefully separate the white, and that is *Sulphur Naturae Jovis*, which put into a little Matrass and fix it, (which may also be done by frequent ⚖mation) make also *Sulphur Naturae* in the same manner, and with the same Menstruum, which dissolve into Oil in B. with which imbibe the said *Sulphur naturae Jovis* until it is fusible, and then it will transmute ♃ into ☽.

Elixir Rubrum.

Take Vitriol of ♀ well purified by solutions and coagulations, unite it with liquor of ☿⚖mate and ✳, then distill a ▽ from it in ashes; then having stood (cold) twenty four hours, distill more ▽ from it. Repeat this until the remaining matter is well broken; then join all the distilled waters to it again, and digest it *in fimo* for forty days: Then distill its spirits, with which imbibe the remaining earth; dry it with a gentle heat, then imbibe again, and dry as before: Repeat this till the earth has imbibed all its ▽. Then distill it, and you shall have a Philosophical ☿, and what sublimes is the Sulphur, which keep apart. Repeat the imbibition and distillation, till no more Sulphur will ascend; with this Sulphur imbibe half its weight of the ☿, put them into a Matrass, which seal hermetically, and fix them together; and this work must be repeated four times, every time with the same proportion of the said Philosophical ☿. Then fix this matter in a vessel sealed hermetically by degrees of △, and all the colours will appear one after another, until it becomes white, and lastly, to an incombustible red.

Take one part of this red powder, cast it upon ten parts of sublimed ☿, set it to putrefy for thirty days, and it will become Oil, which being Projected upon boiling ☿, will transmute it into pure ☉.

The said red powder being infused in wine overnight, and consumed in the morning, cures most diseases in man's body.

The best way to Extract the ☿ of ♂

Sublime flowers of ♂ after Glauber's way, in great quantity, in casting the ♂ in powder upon kindled coals in a furnace with many pots one upon another, wherein the flowers settle. The flowers which are in the last or highest pots must be received into running ☿, by distilling them in a retort with two parts of soot, and one of black soap. Those in the middle, by black soap and salt of Tartar: Those that are in the lowermost pots, by soap only, with a little salt of Tartar, not much, left it should reduce the flowers into Regulus.

The furnace must be round, and well-made everywhere, then set a cover upon it like a funnel, and the pots upon that; then fill the furnace with coals, and let them be well kindled before you cast in the ♂, that the flowers may be pure and white; then cast in the ♂ through a hole, which must be on the side of the cover: And thus you shall Sublime lbj. of flowers in an hour. (see the first Figure)

The process teaches us to set fifteen or sixteen pots one upon another; but I think five or six may do as well.

To Extract ☿ of ☽ or ♄

Dissolve filings of ♄ in A. F. one part, and fair ▽ two parts, precipitate the Calx with salt of Tartar, then add crude Tartar to this Calx, and boil them together a long time; at last, revive it with hot ▽ and you shall have a fluid and running ☿. In the same manner you may proceed with ☽.

TAB I   pag: 72.

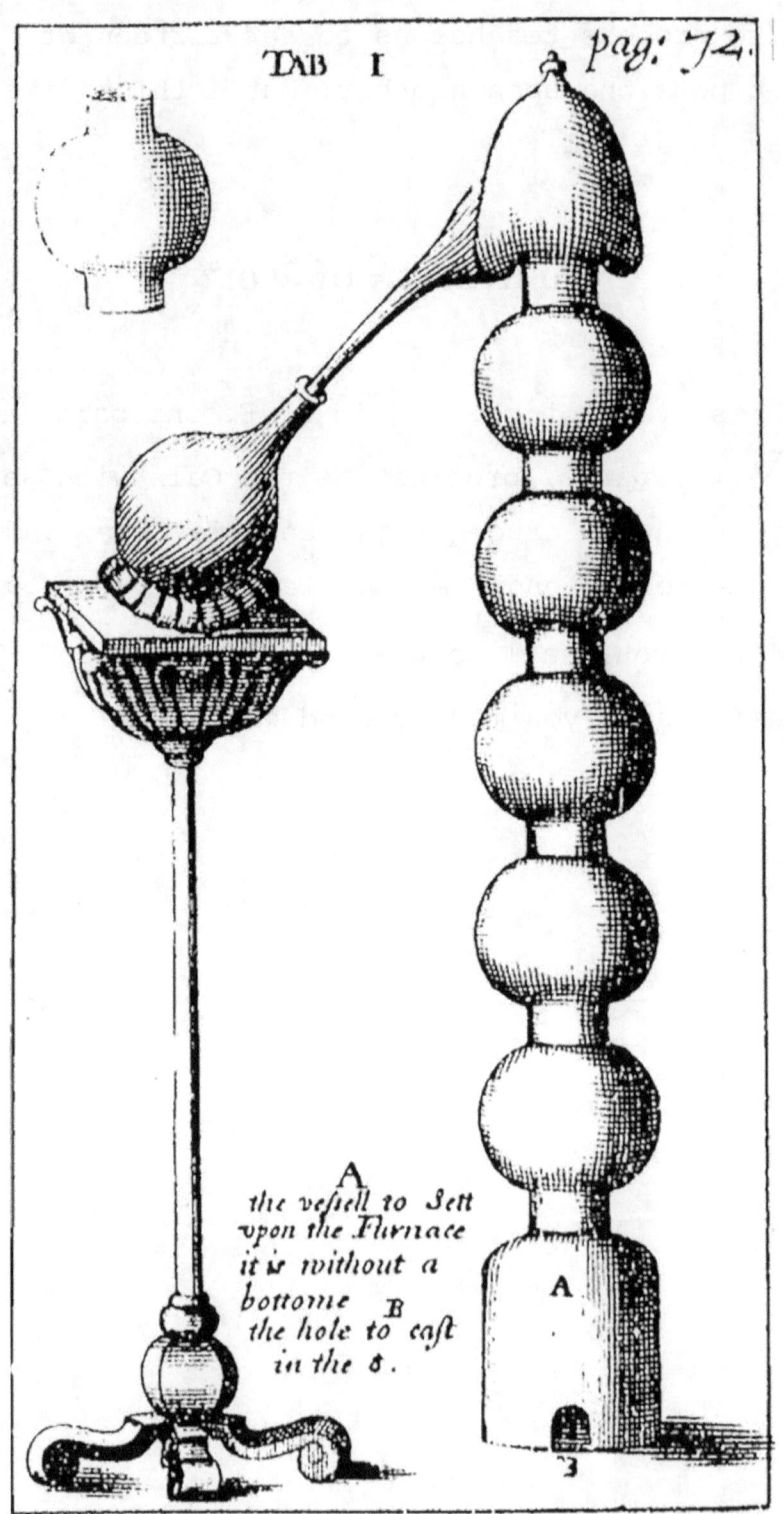

A
the veſſell to ſett
vpon the Furnace
it is without a
bottome B
the hole to caſt
in the ɞ.

To make a Minera of ☿ of Antimony, ad Infinitum.

Take of the ☿ revived from the flowers above-mentioned ℥viij. Sublime it with salt and Vitriol, according to Art: Then take of this ☿ sublimate and ♁ in fine powder, equal parts, mix them well together, and distill a butter thereof, giving gentle fire for four hours, then distill with a very strong fire, and the ☿ will distill in great quantity. Note, that when the butter has come over, before you increase the fire to drive over the ☿, you must change the recipient, putting on another full of water impregnated with ✳. Then take the Cinaber and mix it with black soap and a little salt of tartar; then distill, and shortly you will have near the whole quantity in running ☿. That which remains in the bottom is the true Sulphur of ♁, of which make a Lixivium with ▽, and precipitate the Sulphur.

Another way to Extract the ☿ of Antimony

by an A.R. Given me by Monsieur Carton.

Take Saltpetre of the first boiling without refining it anymore, and Vitriol and ✳, whereof make A. R. after the Dutch way of making A. F. where they put a hundred pounds of matter into a large iron pot with a large stone receiver or a stone pitcher: The junctures all well luted with a lute made of sand, quick-lime and water: They gave △ by degrees, at last very strong; the distillation will be performed in twelve hours. Then let all cool, and take out the A. R. Note, that in distilling this A. R., if your Recipient is not very large, it will be needful to keep wet cloths doubled upon the rec. to abate and condense the violence of the spirits.

Then take ♂ mineral in fine powder, which put into large Jar-glasses, such as they keep thin sweet-meats in, which are as large at the top as at the bottom: You must have many of these glasses, but put not too much into one glass, pour a good quantity of the A. R. above-mentioned upon this ♂; stir and shake them well together by turning the glass in your hands (several times a day) for ten, twelve, or fifteen days, keeping the glasses slightly covered with a wooden cover: And if you put these vessels to digest in some gentle heat, it will be the better.

Note, that the Secret consists in well opening the body of the ☿ by the A. R. and therefore when the time of the digestion is ended, and that you see the ☿ is dissolved, or reduced into a white Calx at the bottom, stir it well together, that the A. R. which is at the top, may become as white as Milk. Then pour off this white liquor (which contains in it the Atoms of the ☿, which are very light, and are easily raised in the A. R. and are well opened) and let it settle, until all the white Atoms are settled to the bottom, and that the A. R. is clear at the top, which decant and put it back upon the ☿ , where you poured it off, which was not dissolved; stir it and digest it as before, then decant the white liquor as before. Repeat this until you have reduced all the ☿ into Atoms or white liquor: Then put all your white liquors and dissolved ☿ together with the A.R. into a retort, and distill first all the A.R. with a gentle heat, until you see the ☿ almost dry, but not hard nor quite dry. Then change the receiver, putting on another large balloon of glass with four or five quarts of ▽ impregnated with lbj. of ✳: Distill by graduated △ for eight hours, then put some coals about the retort upon the sand; give at last very vehement △ above and below, for four hours more, at last, as vehement as possibly may be, and you shall see the rec. filled with white fumes, which will revive into running ☿ in the water in the Recipient, and part of these white fumes will become a thicker substance, like melted ♄, and part like *Mercurius vitae* but of what consistence soever they are, you may easily revive them all into running ☿, by washing them in warm water.

The ☿ of ♄ is made in the same manner, but in greater quantity, taking Ore of ♄ instead of ♂ Mineral.

To Extract the ☿ of ♂, or of ♄.

Wrought several times by Monsieur Van Outre, Physician Of Brussels.

Take Antimony Mineral, (or a Calx of ♄) in subtle powder, imbibe it with spirit of salt until it be like Pap: Digest it seven or eight days, or longer; then distill to dryness: Then change the Recipient, putting on another filled with ▽ impregnated with ✶. Distill it gradually s. a. and you shall have running ☿.

(Sir Kenelm) This extraction is upon the same foundation as that which Monsieur Corton gave me, with an A. R. which he has done often.

Butter of ♂ without Sublimate, to Extract ♀ of ♂

Take ♂ one part, salt decrepitated two parts, and Vitriol calcined to whiteness four parts; reduce all into a fine powder, mix them well together, and cast them by little and little into a retort red-hot through a spout in the upper part of the retort, as Glauber teaches; or distill it in a glass retort luted, in a naked △ and you shall have a butter like unto that which is made of sublimate.

Note, that you may rectify this butter for other operations with soot and coaldust.

To Extract the ♀ of ♂ with this Butter,

proceed thus.

Precipitate this butter with warm ▽, then dry the powder, and mix it with one part of black soap, and two parts of soot; distill in a retort into a Recipient full of ▽ impregnated with ✳, and you

shall have a running ☿, which is the *Sperma* of
of ♀ of ♂.

Another way.

    Take the precipitated powder of the afore-said
butter of ♂, and dry it gently, then mix it with
℥iv. of Tartar, and ℥viij. of Quick-lime, and ℥ij.
of ✳; distill this in a retort s. a.

    Note, that the Calx of ☽, and that of ♄ may be
precipitated with butter of ♂, and then a running ☿
may be distilled from them.

To Extract a ♀ out of ☽

    Dissolve ☽ in A. R. and then precipitate it
with spirit of urine, or with a dissolution of ✳
dissolved in distilled vinegar, and it will attract
what there is of ☿ in the ☽, and the remaining Calx
of ☽ is the running ☿, which is transmuted into ☉
by the Sal Enixe.

Another way to Extract the ☿ of ♂

Take lbj. Of ♂ in subtle powder, boil it in a Lixivium made of Quicklime and Pot-ashes, let it boil for two hours, then let it settle, and decant the clear; then put more Lixivium upon the ♂, boil it as before. Repeat this so often till there be no more Sulphur in the ♂, which you may know by pouring distilled vinegar into the decanted Lixivium; when there precipitates no more Sulphur, or when it changes no more. Then edulcorate well the residue of the ♂, and dry it, then grind it with ℥ iv. of salt of Tartar, and as much of ✳, and ℥viij. of Vitriol; put this to sublime with a gentle △ at first, and at last very strong △ for seven or eight hours, and all will be sublimed.

Make this sublimate, and mix it with an equal weight of Quick-lime, distill it in a retort into a Recipient almost full of ▽ impregnated with ✳, and you shall have a running ☿ of ♂. In the same manner you may extract the ☿ of ♄, taking Calx of ♄ instead of ♂.

Mercury of all Metals.

Take salt of Tartar, and powder of Pebble-stones,[10] mix them well together, and cast them upon burning coals, and there will ascend a spirit, which must be received, which has a virtue to convert the Calx of Metals into running ☿.

A great Secret, ☿ of ♂, and other Metals,

ad infinitum.

Take ☿ of ♂, sublime it with salt and Vitriol in the ordinary manner, without calcining the materials. Take of this ☿ ♎ mate, and ♂ in powder, *ana*; distill a butter thereof: Then take the Cinaber, and grind it with that which rests in the bottom of the retort, and distill a ☿ thereof, which will serve for the like work. Let the butter resolve in the air into a liquor upon a marble stone, or upon glass in a moist place, then pour of this liquor upon ♂ in a cucurbite; digest in sand for two days, then distill it, and there will come over a red Oil with the flegm, which is the Sulphur, or ☿ of ♂, or a natural Oil of ♂: For if you leave this

---

[10] Silex. HWN

menstruum with the Oil in the open air for two or three hours, the Oil will precipitate to the bottom in a red powder, which will burn like common Sulphur. Mix this red powder with two parts of soot, and one of soap, put it in a retort, and distill it, and it wifi revive into running ☿, which will distill into the rec. which must be almost full of ▽ impregnated with ✳.

In the same manner you may also extract ♀ out of other metals, mixing this Oil with their Calx, soot and soap. The said menstruum will serve again, putting it upon new ♂, extracting new Sulphur from the same, or red Oil, which precipitates into Sulphur, as was said, or into running ☿ by revivification. And in this manner you may make a perpetual Minera of ♀ of ♂, and of other Metals, *ad infinitum*.

Note, that other Metals must be in very subtle Calx well opened, that the said menstruum may act the better upon them.

Note, that if the salt of the earth be well extracted, and reduced to a Sal Enixe, wonderful operations may be done with it; and if you take of this Salt Enixe and of Vitriol, and make a sublimate thereof with ☿ of ♂, and then a butter of this ☿ ♎mate and ♂ Mineral, and join this butter with one

113

part of the Lunary Butter, made as was taught, and with that make a dissolution of ☉, you will have an Aurum Potabile and a universal Medicine; and without doubt a powder of projection upon baser metals. This matter is an admirable Chalybs or Magnet of the spirit of the world, being exposed to the open air for some time, and then put in a vessel, and sealed hermetically, and digested for forty days (or fifty) you will see such effects, as will promise a happy success, and yet better, if you add the Sulphur of ☉ drawn with Regulus of ♂; But this requires the conduct of an able operator.

To prepare the Common ☿,

so that it will have all the Qualities and Properties of ☿ of ♂, and will be as Powerful to Volatilize ☉.

Amalgamate lb ij. of ☿ with lb j. of ♃, thus: Melt the ♃ in a crucible, then take it off from the △, and being near ready to congeal, pour the ☿ upon it, and stir them well together with a stick, then cast it into fair ▽. Then with these lb iij. of *àáà* grind lb iij. of filings of ♂, and lb iij. of ♂,

114

and all being well mixt, put it in a retort, and distill over all the ☿ into a Recipient full of ▽.

Another Excellent Preparation and Animation of ♀

Take of Martial Regulus of ♁ ℥iij. and ☉ ℥j. melt them together, and make a Regulus, which pulverize, and then grind it with common ♀, then distill it seven times, and you shall have a ☿ very pure, and fit for any operation.

Another way.

Amalgamate lb ß. Of ♀ with lbj. of ♃ in a crucible, and being cold, grind with it lb ß of filings of ♂. Then distill over the ♀ with a strong △.

Another.

Grind ♀ with ♁, Quick-lime, and Tartar, then distill it.

Another.

Take ☽ one part, Regulus of ♂ two parts, melt them together; then reduce it into very fine powder, which grind and *àààte* with ☿, then distill seven times.

Another.

Join ☿ with Sulphur in the form of a Cinaber, then add salt of Tartar and soap, to the consistence of a paste, whereof make little balls, which distill in a retort.

About a Particular Spirit of Nitre.

It is not a common spirit of Nitre, but it is a spirit, which by many cohobations and distillations renders its own body volatile in the form of snow, which melts with the least heat, and is congealed by cold; and that is that *Acetum acerrimum*, which dissolves all metals, and reduces them to their first matter, and perfect metals being dissolved therein, will be coagulated, and perfectly fixed, which will change other imperfect ones into perfect by projection.

Take ☽ and ☿, q. v. *ááá*te them together, then mix this *ááá* with half its weight of ☿⚖mate, then put it in a retort, and pour upon it of our acid spirit, and the matter being well dissolved therein, distill and cohobate upon the remaining body so often, till all the matter be converted into a volatile spirit, and nothing remains in the bottom; that which does not ascend, must be made volatile: Then dissolve that volatile again in more of our acid spirit, and distill and cohobate so often upon that which remains in the bottom until all be fixed again, and this fixed matter render again volatile, and the volatile fixed again, until it be tingent and penetrating, and be a fusible salt abiding in the △[11].

You must have the spirit of natural fusible salt, which is the principle of all metals, vegetables, and animals; this spirit being purified and reunited with its body (also purified) renders its body volatile and unites itself inseparably with it, and becomes a volatile fusible salt like butter, which congeals being cold: This butter dissolves all metals, as warm ▽ dissolves ice, and is the true matter of the great work, and the Philosophical ☿.

---

[11] This is an error: the symbol should be ▽. -pnw

To prepare the universal spirit, which is the universal salt, you must purify and rectify it well, and by its means, volatilize its fixt body, (also purified.) For to render the fixt volatile, the quantity of the volatile must exceed the fixt; and also to fix the volatile, the quantity of the fixt must exceed the volatile; but the long digestion supplies the quantity of the fixt, because that which is naturally fixt is contained (although changed for the present) in the volatile: But the addition of ☉ (which it dissolves, and unites itself radically with) shortens the time, and hastens the fixation: And then to render it from volatile fixt by a long digestion; when it is volatile, it will pass over in a retort like Oil, which will congeal, being cold, and melt with heat; 'tis the Sperma of metals. For to fix it the better and the sooner, you must add ☉, and digest.

An Operation upon ♄: Sent me by Monsieur Boucaud.

The Philosopher's Epilogue.

Solution and ablution are one and the same thing, for by calcination the body is divided into small parts; by putrefaction it is corrupted, and when it is distilled, it is reduced into its first matter, and remains dissolved.

Congelation is a fixation, re-union, or coagulation of the volatile and dissolved body.

By reduction and fixation, when this body is sublimed, it fattens and resolves, unites, and at last is perfectly coagulated. Thus in these two solutions and coagulations, are contained ablution, reduction, and fixation.

Quintessence of ♄; the Universal Dissolvent.

Distill fifty or sixty quarts of vinegar, and before you distill the vinegar you must evaporate a fourth part of it, which is nothing but flegm; and for to render this distilled vinegar more dissolving, it should be distilled once or twice from Lees.

Take lb xij. or xv. of good English Lithargy of Silver, reduce it into fine powder, and put it into Matrass of three or four quarts apiece, put lb j. into each Matrass; then pour upon the powder so much of the distilled vinegar, as may cover it the breadth of six inches; then put them in digestion, with the second degree of △ for two days, in which time the distilled vinegar will be of a yellow colour, and very sweet. Then decant this distilled vinegar impregnated with the essence of ♄, and put on fresh upon the lithargy; digest as before; then put in all the decanted vinegar and filter it, and distill in several cucurbites with a gentle heat three parts of the distilled vinegar; put the remainder in a cellar, and in twenty-four hours the greater part of it will be congealed into a substance like ice; it will suffice if you distill it off to a syrup: Then upon this syrup pour new distilled vinegar, about the same quantity as before, digest twelve hours; then distill off about the quantity you put on: Put new distilled vinegar upon the residue, somewhat more than the first time, digest and distill as before. Then pour upon it about half the quantity of the said distilled vinegar that you put on before, digest and distill as before. Repeat this digestion for twelve hours, and distillation so often, till you find that the distilled vinegar comes off in the beginning of the distillation as strong as it was before, which is a sign of a perfect attraction of the universal dissolvent made by the distilled vinegar.

Then put your Gums which remained in the cucurbites into one or several large matrasses, which stop and lute well, that nothing may exhale; then put to digest in B. vapor, or *in fimo* (which change every six days) for twenty or thirty days, more or less; for the sign of a sufficient digestion

120

is, when the matter comes to be black, and that it acquires as it were a stinking scent, which is a sign of its mortification, by which it ought to acquire a new life, and a spiritual vesture. Then divide this matter or ceruse into several parts, which put into several retorts, which you may do by causing the matter to melt with some gentle heat, and then pouring it hot into the retorts, for it easily congeals by cold; and if any of it congeals about the necks of the retorts, make it melt, and run down; the retorts must be of such a bigness, that at least four parts of them may remain empty. Then distill off all the flegm with a very gentle heat in sand, and so soon as you perceive any fumes or vapors, cease, and let all cool; then change the Recipient, putting on a large one, and having well luted it, and the lute dry, give the $\triangle$ by degrees, at last very strong and vehement, until you see no more fumes come over, but that an Oil or Gum as red as blood distills over. Take the feces remaining in the retorts (which will look like black ashes) and extract the salt out of them with distilled vinegar, as you did with the lithargy, which salt will be in long rocks like rock-salt-petre; and this salt will be more subtle than the first: Distill this salt in a retort, putting what distills to the first liquor; out of the feces extract again the salt, of which distill also the spirit in a retort. Proceed thus

121

until the remaining earth, or Caput Mortuum gives no more salt. Then take all your spirits, and mix them together, and put it in a large and high cucurbite, which cover with a double paper Oiled and dryed; tie it well about the neck of the cucurbite with a pack-thread, then put on the head, and lute well the junctures, put on a pretty large Recipient, with a narrow and a short neck; distill in B. vapor. and the AEthereal spirit will pass over through the paper, and the flegm will stay behind, because it cannot pass through the Oiled paper; and if your spirit is not subtle enough, you may rectify it once or twice with new Oiled paper; then keep it in a vessel close stopped in a cold place: Then take off the Oiled paper, and distill the rest of the liquor to the consistence of a red syrup; put the cucurbite with the syrup in a cellar, and in two days' time there will be many little crystals very white, which separate, and wash them in the flegm, and they will be white and pure: Then put the flegm to that which remained in the cucurbite, and distill to a syrup, which put in a cellar to crystallize as before.

Cleanse and wash the crystals to whiten them, then put them together upon white paper to dry them for two days in the shadow; then put them in a cucurbite narrow and somewhat high, and pour upon them of the afore-mentioned AEthereal spirit, so much as may cover it the breadth of three or four fingers, digest twenty four hours, then distill in B. M. All the spirit will come over, and in the bottom will remain a clear and transparent gum, upon which pour again the distilled spirit; digest and distill as before. Repeat this cohobation and distillation four times, at the fourth time the said gum will distill over in the form of an Oil as white as snow; swimming upon the spirit: This Oil is the true and only dissolvent of ☉, separate it from the spirit by a funnel. And thus you shall have the Philosophical Menstruum, the vegetable, and mineral salt, Aurora Dianae, and the true Philosophical ☿, and the precious ▽ dissolving the two Luminaries, into a physical dissolution, with which you may prepare medicines both for health, and for projection to transmute metals, which will be both short and easy, as follows.

'Tis not enough to have the Menstruum or Philosophical ▽, for it serves only for an agent or a means to excite the vegetative quality which is hidden and buried in the occult secrets of the Metalline Nature. And it does not suffice only to know that ☉ makes ☉, and ☽, ☽ but it cannot render them apparent, except the said bodies be first discontinued, that is to say, that this Metalline form be reduced into subtle parts attenuated, for to be afterwards opened and reduced into Calx, of which this Menstruum easily draws the fixt grain or Sperma, the principle of vegetation.

Prepare then a slight, spongy, well opened, and attenuated Calx of ☉ which put in a small cucurbite, and pour upon it so much of the afore-mentioned white Oil as will cover it a fingers breadth; digest two or three days with a gentle heat, then distill over all the Oil, then pour the spirit upon the Calx: Then pour upon this matter four or five times as much of the above-mentioned spirit; digest twenty four hours, and the spirit will be tincted of a pure red colour, more red than any Ruby, which decant, and dry the remaining matter, and pour upon it the same Oil, and digest twenty four hours, and it will be very red. Repeat this so long till your ☉ will yield no more tincture. Then circulate all your Tinctures in a Pelican for thirty days, and then separate the clear from an Hypostace which will be at the bottom, and you shall have the true Aurum Potable, which will be of an admirable virtue, taking three or four drops of it at a time in a little sack, or other fit vehicle.

But for the work, you must separate the spirit by distillation in Balneo, until the tincture remain in the bottom in consistence of an Oil, upon which cohobate the spirit, and distill as before: Repeat this seven or eight times, and the said tincture will remain like an Oil that will congeal no more, which is the Philosophical Aurum Potabile, having a vegetative virtue, being sown in its own earth, which is the Calx of ☉, prepared as shall to taught hereafter.

The Philosophical Aqua Regis.

Take Nitre and ✳, *ana* ℥iij. reduce them to fine powder each by itself, then mix them well together, and put it in a retort of three or four quarts, and distill in sand into a very large Recipient, the junctures well luted with paper, and paste made of flower and ▽; for if you should take a stronger lute, all would break: Give the △ by degrees, until you see white fumes in the Recipient; in half an hours' time all will come over; then let it cool, and you will find in the Recipient about ℥j ß. and about the neck of the retort a sublimed salt, which proceeds from the ✳, which will not dissolve except in hot ▽; the retort being cold, take out the Caput Mortuum as well as you can, and the retort being sound, put in fresh matter the same quantity as before; repeat this till you have ▽ enough: Then digest this ▽ in ashes in an alembick with a gentle heat to separate the flegm from it, which will be insipid; then distill the rest in a retort, and keep it for use.

Take ℥j. of ☉ well purified by ♂, beat it into thin plates, cut them small, and put them in a crucible and ignifie them: Put ℥vj. Of ☿ in another crucible, heat it until it begin to smoke, then take it from the △, and pour it upon the ☉, stir it well together with a stick until it be well *àààmated*, then cast this *ààà* into a marble mortar, grind it well, pouring on fair ▽ to wash it from all its blackness and foulness; then squeeze out so much ☿ of this *ààà* as you can: Then grind this *ààà* with equal weight of prepared salt; put it in a retort, and distill over all the ☿ in sand into a Recipient half full of ▽: The ☿ being all over, increase the △ for four hours, so that the bottom of the retort may be always red in the sand; then let all cool, take out the retort and pour hot ▽ into it, and let it stand so for an hour, and the ▽ wiil dissolve the salt; pour it out, and pour more hot ▽ upon the matter; do thus three or four times. Then pour out the ☉ with the ▽ into a Poringer, which will be very subtle powder; dry it gently, and put it in a Matrass, and pour upon it of the above-mentioned A. R. about ℥vj. stop the Matrass with

126

cotton only, and put it to digest in hot ashes, and in a few hours it will all be dissolved into a liquor of an orange colour, leaving some impure earth at the bottom. Upon this dissolution pour of the ☿ which you drew off by distillation about twice the quantity of the ☉, digest it for two or three days, or so long until the ☿ be all dissolved, and the ▽ be clear like rock ▽, and the ☉ be in the form of a light sponge in pieces, swimming in the ▽; separate the ▽, and wash the ☉ with salt ▽ filtered, then wash it in fair ▽ so often till it be well edulcorated, then dry this powder of ☉ and it is prepared. For to attenuate it further, and to render it more spongeous, mix it with double its weight of sublimed ✳, grind them well together, and put them in a small cucurbite with its head, and sublime in sand all the ✳. Then grind this ✳ again with the ☉ and sublime it once more, so will the ☉ be well attenuated and opened, and fit to be joined with the vegetable salt. Then put this powder of ☉ into a Poringer of stone-ware, not glazed, and pour upon it some good Oil of Tartar, dry it gently, and pour more Oil upon the powder, and dry it as before: Repeat this till you have employed ℥iv. of Oil of Tartar to ℥j. of ☉; then put it into a Matrass with a short neck, stop it close, and put it in an iron pot in sand, then cover the pot with any other pot, and give △ of reverberation, so that the ☉ may be red in the Matrass, but not melt; continue the △ in that degree for forty eight hours. Then take out the Matrass, and wash the matter with hot ▽ till the ☉ be well edulcorated, then dry it, and imbibe it again with fresh Oil of Tartar; reverberate it as before for forty eight hours. Repeat this work twice more; and you shall have a very light and spongy Calx of ☉.

(Hartman) Note, that instead of this Calx of ☉, you may take one prepared, by calcining it with flowers of Sulphur, as Sir Kenelm Digby prepared it for Saunier's work, which see in its place.

Then imbibe it once with Oil of Tartar, and proceed in all as before.

Having reduced the ☉ into an Oil, it will be necessary to have an earth of its own nature, to make it grow, and produce the fruit which we expect of it.

Now this Calx of ☉ shall serve for an earth to receive this seed. But since that in all bodies there are three things, to wit, the soul, the spirit, and the body; that which has a body, cannot receive the soul, except it be opened by the spirit: It will then be necessary to reduce the ☉ into a spirit, which is done by reducing it into ☿, its first and nearest matter; which to perform, proceed thus:

Take ℥j. of ☉ well purified by ♁, reduce it into thin plates, cut them small, and put them in a Matrass, pour upon them ℥vi. of our Philosophical ▽, keep it in digestion till the ☉ is all dissolved, then distill off the ▽, which cohobate again, and distill as before. Repeat this three or four times, then distill off about three parts of the ▽, expose the rest with the vessel to the open air, and the ☉ will congeal into crystals, which put in a glass bottle, and stop it very close, and keep them in a dry place until they be dry; then grind them with twice as much ✳ sublimed with salt; put this into a large Matrass, and pour upon it by drops of good Oil of Tartar, the double quantity of the ☉, or until it be of the consistence of thin mustard; then seal it hermetically, and keep it in digestion with a gentle heat for two and forty days, during which time the matter will putrefy and smell very strong, and you shall see all the colours appear successively; take a little of it and wash it well with warm ▽ several times, then being dry, put some of it upon a thin plate red-hot, and if it melts without smoking, it is a sign that it is all

Mercurial, and well prepared; but if it smokes, you must keep it in digestion until that sign appears: Then wash and edulcorate it well from all saltiness, and dry it very gently; then mix it with seven parts of prepared salt, put it in a cucurbite, which put in sand, and give a gentle fire for twelve or fourteen hours, then increase the fire, and continue that degree as long; continue the sublimation until all the Philosophical Calx be sublimed: Then gather carefully with a feather this sublimate, and put it in a glass mortar with warm ▽, grinding it with a glass pestle for an hour or two, then let it settle, and pour off the ▽, put on fresh hot ▽ and grind it until the matter comes to be of the consistence of mustard; then add good white-wine vinegar, and grind it until all is converted into running ☿.

Composition.

Take ℥ß of your Calx of ☉ prepared and attenuated, as was said, put it in a glass mortar, and pour upon it ℥iij. of the Solary ☿; the ☿ will suddenly swallow up its body, as one drop of ▽ mixes itself with another; then squeeze out so much ☿ of this *àà*, that there remains but about two parts of ☿ with the ☉. Put this *àà* in a

Philosophical egg, and pour upon it by little and little of your Oil of ☉ before-mentioned, hold it over a gentle △, and stir the matter with an iron rod, that all may well mix and incorporate, pouring on so much of the said Oil, that it be of the consistence of thin mustard, and then you shall suddenly see marvelous things, when the soul of the said ☉ (which is its Oil) enters into the body of the ☉, by means of the spirit, which is the solary ☿, and that by means of the said soul, the spirit unites with its body, of three being made one; stop the vessel speedily, because of the fumes. The body of the ☉ which was dead before, being by this only and admirable means animated, dignified, and filled with a vegetative life, will thereby acquire an inward power of multiplication, as well as the sperms and seeds of all animals and vegetables, and be made fit to grow and produce fruit, (being sowed in a fit earth) which it could not do before, because of that default. The vessel being sealed hermetically, put it in ashes in a brass vessel in the shape of half a boul; digest it with a Lamp △. As for the time, and the colours, mark what Trevesan said of it; for at the end of forty days you shall see the blackness: Continue the first degree of heat to whiteness, which will appear within four months: Then augment the △, and continue until it comes to

131

be of a citrine colour, and then there will be no more danger. Increase the △ to the fourth degree, and continue that, till your King takes his robe, and that the matter suffers ignition without smoking.

(Hartman) This process was sent to Sir K.D. in a letter from Paris, by Abbot Boucaud, with the following words. Sir, I have sent you here enclosed a great work upon ♄, which Monsieur de Rouviere has given me; it comes from a man who having been carried away, and kept close in a castle, made at last his escape, and was conducted to the Duke D' Elboeuf, and Monsieur de Rouviere found the said process under his Boulster.

The said Abbot sent to Sir K. D. also the following process, which he said he had from an intimate friend, who assured him that it was a reality.

Take of a very good ore of ♄, that was never wet, or instead of it, take a true and natural mineral lithargy, not artificial; pulverize it and grind it upon a marble stone with distilled ▽ several times distilled: Put your ore of ♄, or lithargy in one or more cucurbites, and pour upon it of the aforesaid distilled ▽, or distilled dew, so much as may cover it the breadth of seven or eight fingers, cover it with a blind head, and lute well

132

all the junctures, and digest for forty days with a gentle heat, shaking the vessel often; when you perceive that the Menstruum is coloured, decant the clear, and put on fresh ▽, or take new ore or lithargy and extract as before; filter the Menstruum and distill it with a very gentle heat. Take this salt of ♄ and put it in a Matrass, digest it with a Lamp △ with six small wicks, and it will dissolve of itself and there will settle to the bottom some impurity or feces; break the Matrass (being cold) and take the pure part and put it into another Matrass, dissolve it by digestion as at first, separate the pure from the impure. Repeat so often till this salt leaves no more impurity. Then keep it carefully, until you employ it in the following work.

Take of this salt ten parts, and one part of ☉ mineral that has not been melted, put them together in a Matrass, seal it hermetically, and digest with a very gentle heat, and there will loosen itself from the salt of ♄ some spirits, which by falling down again will dissolve the ☉ by little and little, and there will separate itself yet some feces which are not useful for this work, which you must separate. Take what is clear and transparent, and put it in a Philosophical egg, seal it hermetically, and digest it with a Lamp △ with a gentle heat, continue the digestion without ever increasing the heat, etc. The said Abbot said, that this was all he could have, or know of this work hitherto.

The said Abbot sent also the following process in a letter from Paris.

Monsieur de R's. operation, by which he fixed ☿ into ☽ with the salt of Saturn and ☽, is thus. He took one part of ☽, and three parts of ♀ whereof he made an *àà̀à* which he put in a Matrass, and put on it *Saccharum Saturni* (made the common way) about a fingers breadth over the *àà̀à*; then he sealed the vessel, and digested it with a Lamp △ with gentle heat, increasing the heat by degrees, it passed through all the colours; and of one Marc of ☽ and three Marcs of ♀, there remained ℥ xij. of fixed matter that suffered fusion and the test.

He said that there should have remained one Marc of the ♀ fixed, but the operation was not well wrought. In another letter he said, that there remained ℥ iij. or iv. of ♀ fixed into ☽, which endured fusion and the test.

134

An Operation upon Jupiter.

Distill a Menstruum out of Vitriol and ✳, with which make Sulphur Naturae Jovis: Make also with the same Menstruum Sulphur nat. ☽ which dissolve into Oil, and with it insere *Sulphur Jovie ad fusibilitatem*, and then project upon Jupiter.

Dunston thus: Having taken our white earth, you may putrefy by itself, or with the Calx of other metals, and change its colour into a new white or red: Then ferment it with the Oil of ☉ or ☽, &c.

Ripley (in his *Viaticum*) thus: Calcine ♓ into a most subtle Calx (for in it there is pure ☿, not brought to its full perfection by Nature) which is easily hardened with the Oil of ☽. Do your work therefore with Tin (until you are rich) because so the work is easily done, and at small charge.

Lullius (in his *Magia Naturalis*) thus: Make *Sulphur Naturae* (without which nothing can be done) and thus of any metal (which he directed to do in a very tedious way) then incere it with Oil of Ferment (as in his Pract. Brev, or Sermocinal) until it is fluid; then is it a perfect medicament.

☉ & ☽ ex ♃

Take of the filings of ♃ lb j. salt-petre lb j., mix them, and separate the spirit from the anima by combustion, subliming it in so many pots as you know: Dissolve the Caput Mortuum (which will be fixt as a stone, so that you may strike △ out of it by collision) with ☿, that there may be a Regulus made of it, which pour out and make into rods, and cement them with store[12] of Calx-vive on a Circulary △, then Coppel them with Lead, and add to them fine ☽, what then remains upon the Coppel is good, and you will have considerable gain by it, and by the separation of A. F. you will have three parts of ☽, and one of ☉.

But when you melt your Caput Mortuum of ♃ with ♂ into a Regulus, as before, when you have precipitated them with Tartar, or mixed them, then put your Regulus to Coppel, and in it you will find ☉: See that you do not cast away the Scoria, for you will find Silver amongst it; Coppel it therefore by itself with the following powder, so you will find ☽, which separate with A. F. (the powder make thus.)

---

[12] "store" is the word printed in the text of 1682. -pnw

Take Chelamus (Bay Salt) melt, dissolve, filter, and coagulate it; melt it again, and do this work thrice: Then cast in this salt into the aforesaid Scoria (from which you separated the Regulus) after you have put it to Coppel, so your work will be done and accomplished speedily, and with great fruit and profit in the applying the fire.

(Hartman) The famous Tachenius relates (speaking of the malignity of Arsenic) that there are some who can burn pure Tin into powder, which cannot again be reduced into Tin by Vulgar Art, as other metals; yet with Arsenic it is made Scoria, part of which by a singular skill becomes pure ☽. Sigismund Wan, a citizen in Voitland, knew and practiced this Art of Separation, to his great benefit; for in the year 1464, he built and endowed a great hospital there, which, as Gaspar Bruschius relates, is at this day to be seen, with the epitaph of the aforesaid citizen.

Now, that ☽ may be gotten out of Tin with Arsenic, Clavious proves in his apology against Erastus Second Vol. Theatrum Chym. Fol. 39.

A worthy gentleman lately related to me, that he knew one, who told him, that out of lb ß. of Block-Tin he got so much ☉ as he sold for 3s. 6d.

137

A short and clear Description of
The Great Philosophick Stone.

The First Operation.

Take salt prepared, Nitre, and Roman Vitriol, *ana* lb ij. beat them into a small powder, mix them, and put them into a pot upon a slow fire, and moving them, cause them to melt, that they may be dryed a little. Then take ☿ taken out of the Mineral, lbj. which being put into a linen cloth, squeeze it and pour it upon the hot matter, moving it with a rod, until the Mercury is hid in the matter; incorporate the mass well when it is cooled, in a marble mortar; then dry it all in a pot very slowly, until it be so dry, that a sword held over the pot, receives no moisture from it; then put it into a sublimatory, and sublime it first twelve hours, afterwards increase the fire, that all the Mercury may be well sublimed, white as snow: So the ☿ lacking nothing of its weight, will be associated with the invisible Sulphur of Vitriol, and purged from the earth and its blackness; and if you will experiment that conjunction, you may separate Sulphur of Vitriol from ☿ thus: Take distilled vinegar, q.v. quench burning iron several times in it, let your sublimate stand therein all night, afterwards pass it three times through a filter, then set it upon a slow fire; so a black scum will swim above the vinegar, which take off; then evaporate all the vinegar with a slow fire, so you will have an excellent sulphur of Vitriol, and the ☿ will remain by itself in the bottom.

The Second Operation.

This teaches to extract the Quintessence from this ☿ sublimate thus: Make A. F. as follows; Take salt-petre and Vitriol, *ana* lb j. beat them, then mix them together and distill them with a slow fire in a glazed Alembick on ashes for eighteen hours, so that nothing more can distill (but lute all so well, that nothing exhales;) after the above-said eighteen hours increase the fire by degrees, that the water may be distilled, and then continue the same degree of heat until it begins to cease to distill, and so proceed by degrees, until nothing more can come from it: Let the vessels cool of themselves, and seal the Recipient with gummed wax; and when you have put your sublimate, being well beat, in a strong Matrass, pour upon it of this water to the height of one or two fingers breadth, and immediately obturate it well: Set the Matrass in ashes on a slow fire for the space of twenty four hours; and if it be not then dissolved, if you pour ℥vij. of water upon it, add of ✳ well beaten ℥j. or more; close it up presently, and set it upon ashes, so it will be dissolved: It is a very great secret. Then abstract all the water, (the junctures of the vessel being carefully stopped) by distillation on a slow fire of coals, even to dryness: Afterwards taking off the cap, cover it straight with a glazen Operculum, lute it, and when you have increased the fire, the Quintessence of Mercury and Vitriol will ascend at the sides of the vessel; at last make the fire yet stronger, that all the Quintessence may be well sublimed, which, when the vessel is cooled; keep carefully: Beat the black feces, and sublime it once more, if perhaps any of the Quintessence remains still amongst it, so you shall have purged the Mercury, and imbibed more of the spirits of Vitriol existing in A. F. Then dissolve and sublime the matter (that is sublimed) twice more after the same manner, that no impurity may be left in it, so it will be whiter than snow.

The Third Operation.

Beat this matter, and put it into an Earthen urinal well glazed within, which cover with an Earthen glazed head like to a *Paradi*, that they may be exactly joined together; lute the junctures well, and digest it eight days or more in an Athanor on a

slow fire of coals; for otherwise it could not be
dissolved into water.

The Fourth Operation.

Put your matter so dissolved into a Matrass,
close it, and dissolve it in A. B. into water with a
continual slow heat: Distill this water in a little
Alembick on ashes with a slow fire of a Lamp, and
water of Paradise will be distilled, (of which alone
the stone may be made by the method after described)
one drop of which poured upon a plate of any red-hot
metal will thoroughly whiten it within and without;
(Note, that the like may be done with the Lunary
made of ☽ and ♓ if they are poured on a plate of
♀.) After the water is distilled, some feces will
remain, which contain in them earth, air, and fire,
which you may thus separate one from another: Beat
those feces, and digest them in an Athanor, as you
did the Quintessence before, afterwards dissolve
them the same way in M. B. At last distill with a
very strong fire in M. B. by an Alembick, a white
Oil, which is called air, which when it ceases to
drop any more, take off the Recipient, and close
well the nose of the Alembick, and so let it cool of
itself: Then set the Alembick with a new Recipient
on ashes, and draw off the red Oil (which is called

fire) with a strong fire. Cast away the earth that remains.

The Fifth Operation.

If you would make a stone of paradise alone, or Virgin's Milk, you need not separate the elements; but if you have separated them, do thus: Take of $\triangle$[13] or red Oil part j. of air, or white Oil parts iv. and of virgins milk parts viij. Put them together in a Matrass with a short and narrow neck; in two other matrasses put of virgins milk, q. v. Seal them all hermetically, and so put them in an Athanor on a slow fire of small coals burnt till they have ceased from flaming, and so let them stand till by several colours they attain a perfect white: Then (if you will have the stone white) you may take out one Matrass, leaving the other two (if you operate with three at once). Then increase the fire sensibly, because your work cannot easily be marred, and so proceed by degrees, until the matter is perfectly red.

---

[13] This is the symbol that Hans used; verified in the printed text of 1682. -pnw

The Sixth Operation.

Is projection. Take an hundred parts of ☿ purged the common way, heat them in a crucible, and add to them part j. of the white or red stone, and the whole will be a medicine. Then take of this medicine part j. and project it upon another hundred parts of ♀ moving it with a rod, and afterwards melting it. Further, project of this, part j. in a hundred parts, and the whole will be ☉ or ☽, according as the stone is which you took: If you will project upon other metals, melt them, or make them into thin plates, and on them when they are very hot project part j. upon a hundred, and set the plates in the △ for some time. If you would augment the virtue of the stone ad infinitum, dissolve it as often as you please in B. M. and coagulate it slowly in ashes; let Jupiter and Saturn be melted.

The Seventh Operation.

Is the magistery of an Athanor, which build thus: Take of magisterial lute, potters earth, horse-dung, paper carminated, hairs cut, make them all into a paste with salt water and vinegar, and with that paste build your furnace; make a round wall four fingers breadth in height, with chimneys, set upon that wall an iron plate which has four supports, by which it may stay upon the wall, leave some distance between the interior sides of the wall and the exterior of the plate, that the heat may ascend by it; then raise the wall to the height of five inches: Then make an Earthen cap, which on the one side must have a window and convenient door, by which you may feel the heat putting in your hand at it, (which heat must be so moderate and uniform, that you may endure to hold your hand in the Athanor as long as you please); the cap must be carefully luted within and without, and set upon the furnace and agglutinated with clay: Afterwards, when the furnace is sufficiently dried, set upon it glass goblets, and set your matrasses upon them. Mind well all these things, and consider the figures cut in the page over against this. Now, by the holes that are made betwixt the plate and the wall, you may increase or diminish the heat at your pleasure. But note, that upon the goblets, above the Tripus and the plate, you may set an Earthen trencher, and set thereon an egg, which cover with another trencher, so that these two trenchers joined together may be lifted up in the air, and the egg not touch the sides of the trenchers.

A Note from One that Wrought the Stone.

I have had certain notice of one that made the philosophers stone with Leaf-gold and a clear ▽, that looked like rock ▽ but smelled strong. He who wrought it for him (that is, attended the Lamp) said, he had made his liquor thrice before it would dissolve the ☉. The last dissolved it by little and little, it became a yellow Aureal Liquor, then thickened by little and little, at length became a black thick broth, in the end like melted pitch: then changed several colours, every one sparkling like Oriental precious stones, and sparked like fire or stars rose in the glass (which was a large egg sealed hermetically) then fell down again. It was digested in ashes made of old bark of Oak burned (unwashed) and the author said, no other ashes would serve. And the heat was never greater, than that he could endure the back of his hand upon the ashes, which was caused by a Lamp.

(Hartman) This relation is of Sir K. D.

Lauremberg's Observations upon Angelus Sala

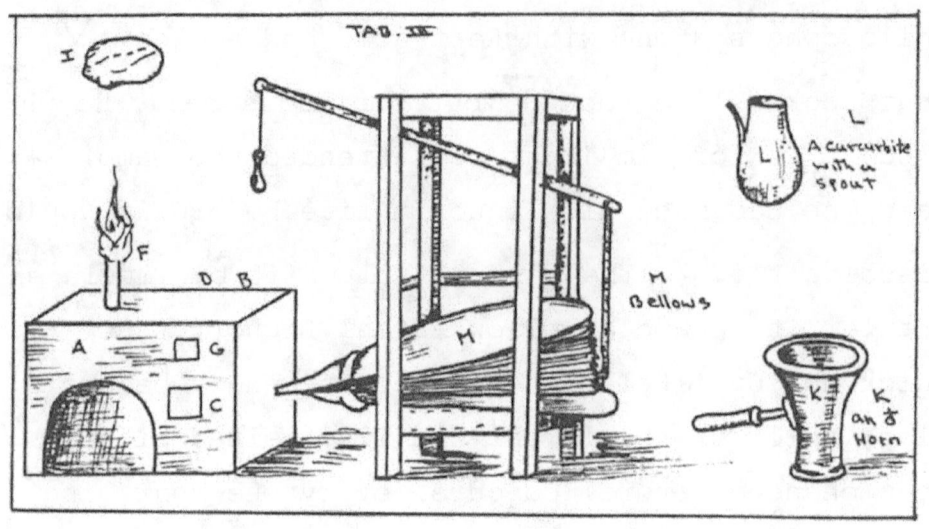

He said: I prepared fluid ☿ such that without the mixture of anything whatsoever imaginable, without any dissolving Menstruum, it did acquire the form of a most pure and transparent liquor; neither hitherto has it lost this liquid form, but is so liquid, that you would imagine it had been brought from a fountain, and which you will more admire, being tasted, it is void of all acrimony, and merely insipid; (I add also this) that some months ago I reduced English ♈ into a fluid and moist liquor, without the least addition of Menstruum, which humidity it not only continually keeps entirely to this day, but (so far as I can see) will never lose it.

(Afterwards in the same page, he said,)

I confess ingenuously, that not long ago, I had the happiness of seeing at a friends, and feeling such an unfactitious liquor (Liquorem ατπιϲψ)[14] whereby leaves of Gold and Silver were dissolved into a pliant and fluid liquor, without any noise or the least suspition of Acrimony. This liquor can be no other than congealed air, without which the life of animals becomes no life; and there is no body under the sun in that three-fold Kingdom that is destitute of it, I would rather search its medicinal power with silent speculation, than weary people with tedious and fruitless discourses.

Concerning May Dew.

May dew is the true Minera of the dissolvent. *Aug*. This liquor is such, that if it be gathered at a certain season, two uses, etc. One, that hereby you may infuse Gold in a liquor of its proper seed, when you begin first to dissolve it, etc. *Cosmop*. But this ∇ is said to be the Menstruum of the

---

[14] This appears to be Greek, possibly Υγρό. The printed version of the 1682 text is not as clear as the one referenced by Hans. -pnw

world: Speaking of the element of ∇, the Menstruum of the world is tripartite, etc. the more pure resolved into air. There is in the air an occult nourishment of life, which we call dew in the night, and in the day-time ∇, rarefied, whose invisible congealed spirit is of more value than the whole earth, *Idem*. The principal matter of metals is the humidity of the air (the aerial substance) mixt with heat, ☿ prepared aforesaid is governed by the rays of the ☉ and ☽ prepared in the sea; not one place or one country will afford it you. Experience testifies, that ☉ is not sought for save in mountains, because it can be seldom had in a plain.

Flamel, Artefius, Pontanus, Zacaire, &c. Their Arcanum.

This stone is that about which the above-named authors employed themselves: It is composed of the mineral gluten, made of ☿ and ♂ mineral, by the addition of the solary ferment thus: Distill the volatile crystals or butter from ☿⏚mate and ♂, equal parts: Or distill, or dissolve common ☿ in A. F. Precipitate with salt-water, and you shall have a very white Calx, which dry, and join it with as much of Calx of ♓, and distill the volatile crystals thereof. These crystals are the magnet, by means of which the universal form, or spirit of the world is attracted; which specifies and determines itself in this matter, by resolution in the air in ♈, ♉ and ♊. Put this liquor in a cucurbite, and digest for thirty days with a very gentle heat with a Lamp, to the end, that there may be a natural distillation made of the attracted spirit, which will begin to come over invisibly the first or second day, with the *Idea* of that which it draws, to wit, the ♄, and of ☿, or with a universal mineral form, tending to metallick.

This liquor will continue coming over even unto the end of fifty days; let not the heat exceed that of one's hand. This AEthereal ▽ is the ▽ of paradise, or the mineral *astrum* of Flamel's two dragons, the one is volatile (which is ☿) and the other rampant, which is the ♄, which do not suffer to be touched nor attacked, until their venemous scum (that is to say, the butter) has produced the spirit of the Mercurial-wind, and the scum of the Red Sea. Note, that within fifteen days this sea or butter comes to be very red, with a gentle heat of a Lamp △ in ashes; and this is Flamel's Red Sea. This AEthereal ▽ penetrates all metalline bodies, (being luminated and made red-hot) and tinges them into ☽.

Two drops of this ▽ being dissolved in ℥iv. of spirit of wine, made a virginal milk, whereof the dose is a spoonful: It is a very gentle emetick, because of its crudity or rather Mercurial substance, whereof the virtue operates upwards, because it is moist and airy. It cures the epilepsie by the character which is imprinted into it, by the word *Fiat*, and all astral diseases, as far as humane disposition permits. This is the Coelestial ▽ which does not wet ones hands after its preparation; 'tis the ♀ of ☿, the ▽ or center of the heart of ☿, and the true extent of ♂, but it requires more work.

Take this ▽ (you must have a pretty great quantity of it, and therefore you must have ten, fifteen, or twenty pounds of volatile crystals) and put it in a cucurbite, and with a very gentle heat of a Lamp distill off all the waterish moistness, which by carelessness it might have contracted from the air: There will remain in the bottom a gum, a syrup, a viscous ▽ a radical mineral moisture, which is the eagles gluten above-mentioned, which did fly the space of fifty days continually; by means of this gentle heat, the gluten flies no more, but is the flying *Crapant* and *Zacair's Mercurial* ▽ which is congealed by cold, and liquefied by heat. The authors before mentioned have digested this gluten per se in a Matrass hermetically sealed, without the addition of a solary ferment; but afterwards they were forced to ferment the powder which they made of it. To shorten the work, take seven parts, or nine or ten, or more of this gluten, unto which by heat join one part of ☉ in leaf, or Sulphur of ☉ prepared by the salt *Enixe* (which is best) and digest in an Athanor, or in Flamels furnace (which is very easy) until all the matter has passed through all the due colours, and come to be of a purple citrine colour; then have you the metalline salt, the most high tincture, a treacle made of venom, a most excellent medicine is multiplied in quantity by new addition of the afore-mentioned gluten; in quality, by dissolving *in humido* into a liquor, and purifying by digestion, and then by fixation; Experience will teach other things far better. This method, although it differs much from that of the greatest philosophers, as Lully, Trevesan, Cosmopolite, & etc. (and being but particular in comparison of that high Generalissima) nevertheless it seems to be universal in regard of metals and minerals. Note, that you may also extract a white and red Oil of that which remains, as was said before, and make a new aurifique stone thereof, which those authors have not understood, or if they have understood it, they have not spoken of it. Note also, that this mineral ▽ of paradise, is the philosophers live ☉, and the ☿ of the wise, but not the Generalissima: And this ▽ will serve against all maladies, for it drives them out, according to the intention and inclination of Nature.

To prepare a Ferment or Sulphur of ☉

Make an *àáà* of ☉ or ☽, and ☿; grind this *àáà*, then squeeze it through a leather, the globe remaining in the leather, you must grind again, and then put it in a Poringer, covered with another Poringer, and lute them well together, then put them to a gentle △ for half an hour: Then grind it again, and digest it between the two poringers as before. Repeat this so often till the ☉ or ☽ be in powder impalpable; then incorporate this powder with fresh ☿, grind them together, and digest with a gentle △ so that little or nothing may sublime, and if anything sublimes, put it again to that which remains in the bottom. Repeat this last operation (adding new ☿, grind and digest as before) so often, till the whole body of ☉ or ☽ be converted into running ☿, and that all may be squeezed through the leather: Then put this ☿ animated into a Poringer, which cover, and digest with a gentle heat, so that

152

nothing sublimes; continue the digestion so long, until you see a thin skin swimming upon the matter, which take off carefully (it will be of the colour of ☉ or ☽) put the matter upon the △ again, increase the heat a little, taking off the thin skin as it rises; continuing so long until the matter produce no more thereof: And thus you shall have the Sulphur of ☉ or ☽.

An Operation that Monsieur De l'Oberye wrote

from Monsieur John's Mouth.

Take the Mother-liquor of salt-petre, let it run cold through washed sand, then filter it by Languetes, then through gray paper: Then evaporate with very gentle heat, putting down the skins as they rise upon the liquor; the remaining salt being dry, grind it, and put it to resolve into liquor in a cellar, then filter and evaporate as before. Repeat this purification five or six times, or so often, till it leave no more feces in the filter. If you take lb x. of this liquor, you shall have but lb ij. ℥viij. of purified salt: Of this lbij. ℥viij. you shall have ℥x. of spirit, by distilling it *per se* in retorts in sand; you must put but lb ẞ. of this salt into each retort; deflegm it in B. Take the Caput Mortuum, and grind it, and dissolve it in a cellar; filter, and congeal, repeating this two or three times. Then being very dry, join ℥iij. of it with ℥j. of the rectified spirit; digest and circulate eight days with gentle heat in ashes, and all will be a ▽ of the colour of amber. Put one part of ☉ into ten parts of this liquor, and it will dissolve it (cold) in less than a quarter of an hour: Decant the dissolution when it is clear; one drop thereof taken in a little broth, is a great corroborant.

Put ☿ revived from Cinaber into the dissolution of ☉, and it will become like a gum, decant the clear, and put the ☿ to dry, and it will become hard; melt it between two beds of calcined egg shells in a crucible, and you shall have good ☉.

Venus into ☽:

Sent me by Monsieur de Beaulieu.

Take fixt Arsenic ℥viij. fixt Nitre ℥iv. Oil of Tartar prepared, as shall be taught hereafter, ℥xij. ✳fixt, ℥xv. Let them all resolve into liquors in a moist place every one by itself; then take these liquors and mix them together, and filter them, then put to them ℥iij. of Oil of ☿, and ℥viij. of ☽ (prepared and dissolved in the liquor of fixt ✶ and fixt Sulphur) mix all well together, and put it in a Matrass, and digest *in fimo* for forty days, changing the dung every eight days: Then decant the clear, and the feces remaining in the bottom, dissolve in the liquor of fixt ✱, and put it to the rest of your liquors; filter it three or four times: Then distill it in a cucurbite with a gentle △ in B. M. (not boiling) distill to dryness, and you shall have a white matter like a stone, and clear like a pearl: And to know whether it is perfected, put a little of it upon a red-hot plate of copper, and if it melts like wax, and penetrates through the plate without smoking, leaving the said plate white where it has touched, it is a sign of an entire perfection; but if you find that it is not yet fusible, and that it smokes yet, grind it upon a stone with a pint of ▽ distified from whites of eggs, and distilled three times upon lbj. of Quicklime, grind it with the said ▽, until it be of the consistence of pap; then put to it four times its weight of liquor of fixt ✱; digest *in fimo* for eight days only, then congeal it as before, so is it perfect. Project ℥j. of this matter upon lbv. of prepared ♀, and sometime after cast a little piece of wax into it, at three or four

155

times; then cover the crucible, and leave it in
fusion for some hours.

The Multiplication.

Dissolve $\text{\reflectbox{3}}$viij. of this matter in lbj. of the
∇ of whites of eggs, then add $\text{\reflectbox{3}}$iv. of liquor of
fixt Arsenic, digest *in fimo* for fifteen days; then
distill and congeal it as before, so is it
multiplied. If you reiterate this multiplication
several times, the matter will remain in a liquor,
which will project upon great quantity of Venus.

To fix the ✳ for this Work.

Take lb j. of ✳ in small pieces about the
bigness of a wall-nut; make a paste with Quick-lime
and whites of eggs, with which endow the said pieces
of ✳, let them dry, then stratify them in a
crucible with powder of Quick-lime, let the beds of
the Quick-lime be about a fingers thick; then put
the crucible to a circulary $\triangle$, which increase and
approach once in a quarter of an hour, at last,
cover it with coals, and let it stand so for half an
hour: Then take out the ✳ (the crucible being cold)

156

and wipe off the powder of Quick-lime, then dissolve the ✳ in fair ▽, filter and congeal it; dissolve it in a cellar into an Oil, which keep for use.

To fix the Arsenic.

Take equal parts of Arsenic and Nitre, grind them, then mix them together; fill a crucible half full with this powder, fill it up with salt of Tartar; cover this crucible with another that has a little hole in the bottom, lute them, and set them in a circulatory △, the △ being half a foot's distance from the crucible; increase and approach the △ once in half an hour about two inches, and when you perceive no more smoke come from the matter through the little hole, put the △ close to the crucible, and at last cover it with coals, and keep it so covered for twelve hours, then let it cool, and grind it, then dissolve it in a cellar, and keep the liquor in a glass close stopped.

To fix the Sulphur for this Work.

Take ℥v. of Quick-lime, slacken it in six quarts of fair ▽, and having stood twenty four hours, filter it, and put it in a kettle; then take ℥viij. of flowers of Sulphur, tie it up in a linnen bag, which hand in the water in the kettle, make it boil for an hour, and you shall have a Sulphur incombustible.

Oil of ☿

Take ℥iv. of sublimate in fine powder, put it in a crucible, and pour upon it lbj. of fine ♃ melted, stir it well together; then being cold, put it upon a clean iron plate in a cellar, and you shall have an Oil or liquor.

To prepare the ♀ for this Work.

Take Arsenic one part, decrepitated salt two parts, pulverize and mix them together; then stratify with this powder some thin plates of ♀, cement them for two days, then put them to a strong △ for six hours; then wash these plates from the salts, and beat them to powder, wash the powder with vinegar, and then with ∇ two or three times; then with soap make a paste thereof, which put in a crucible, that has a hole in the bottom, put this crucible in another crucible, and so melt down the powder of ♀, and it will run through the hole into the other crucible, and you shall have a very white ♀, and well prepared for projection.

To prepare the Salt of Tartar for this Work.

Take equal parts of Tartar and Quick-lime, powder them, and mix them well together, put this in a por., cover it close, and put it in a Potters Oven when he burns his pots; then make a Lixivium of it with rain ∇, which filter and evaporate to dryness, mix this salt again with the same quantity of Quick-lime, and calcine it in a potters oven as before. Repeat this five or six times; then dissolve this salt in distilled vinegar, distill and cohobate so often, till it will no more congeal into a salt, but that it remains like melted wax in the bottom, which pour out, and keep it for use.

To Prepare the ☽ for This Work.

Dissolve ℥viij. Of ☽ in ℥viij. of spirit of Nitre, then precipitate with salt ▽, the powder of ☽ being settled, and the ▽ clear, decant; then edulcorate the powder and dry it, then dissolve it again in spirit of Nitre as before; precipitate, edulcorate, and dry the ☽ as before. Repeat once more, three times in all, then put it in a Matrass, and digest it eight days in sand. So is it prepared, and fit to be further prepared, and dissolved in the Oil of fixt ✳, and fixt sulphur.

Transmutation of ☿ into a Regulus.

Precipitate butter of ♂ with warm ▽ once, without further edulcoration; dry it gently, then add a fourth part of ☿, and of black soap, and salt of Tartar, of each as much as needed to make a paste, whereof make little bullets, which put in a retort well luted; distill in a naked △, with a strong sudden △, and the matter being melted, you shall have a Regulus as white as ☽, which must be melted three or four times to have it finer and whiter.

Calx of ☉

Monsieur Le Febore shewed me a very subtle and spongy Calx of ☉, he had made thus: Purify ☉ to its greatest height, beat it very thin, and cut it into small pieces, heat one part of them in a crucible, and fix parts of cleansed ☿ in another: Make an *ááá* in due manner, stirring a while with an iron rod, then throw it into cold ▽; squeeze out as much ☿ as you can through Chambo-leather: To the remaining globe put double as much flowers of Sulphur, grind them well together: Put this mixture into a capacious crucible, and gently burn away the Sulphur, and evaporate the ☿, reverberating the Calx three or four hours after all is gone away. Repeat all this work twenty or thirty times; then reverberate it under a muffle with so gentle heat, that it melt not, the longer the better:

Then burn S.V. three or four times from its *Quære*[15], Of grinding the Calx long with pure virgin honey, then dissolving it in a large quantity of pure distilled warm ▽ and letting it stand warm till all the Calx is settled to the bottom: Also of grinding the Calx with purified salt of Tartar, then reverberating the mass, and lastly, dissolving the salts in warm ▽ and letting the Calx settle, as

---

[15] From the 1682 text. In Latin, this means "inquire" -pnw

with the honey: I think it will be a very subtle Calx, to dissolve the ☉ in A. R. of Nitre and ✳; then precipitating it with Spirit of Urine, or with a marinated ▽ made by dissolving the fixed salt of Urine in pure distilled rain or spring water.

A pretty Curiosity,

To make Metals Vegetate visibly.

Calcine white and transparent River Pebbles, extinguish them in ▽ to have a Calx thereof, which reduce into subtle powder with equal parts of Tartar and Nitre (fulminated together) taking double quantity of this fusible salt: Dissolve this matter upon a marble stone or glass in a moist place, and you shall have a liquor, which filter: Take about ℥ ij. of this liquor, put it in a Viol., and put into it about ℥ij. (or less) of the Calx of any metal, dissolved in its Acid Menstruum. Then evaporate to the consistency of a Calx: Let it stand, and as soon as it is cold you shall see the metal vegetate, and shoot out into branches, which will be of different colours if you put in the Calx of divers metals: This is fine and pleasant to behold. Note, that it is to be observed in general, that the cause of vegetation is the encountering of an airy acid with a fixt Alkali; and it is thus, that Quick-lime calcined with common salt into an Alkali, being spread upon barren ground, fattens it, and makes it fruitful, causing vegetables to grow, by contracting the acid of the air and its volatile salt.

To Engender Cray-Fishes.

It is to be observed first, that to do this operation well, you must do it at the increase of the ☽, and in the sign of Cancer, if possible, or at least in any other aquatic sign.

Then take a parcel of the said Cray-fishes, taken in Brooks or small Rivers, being all alive; divide them into two parts, put one part thereof into an Earthen pot not glazed, cover it with its cover, or with another pot, lute them well, and put them to calcine for seven or eight hours with a strong △, until they be well calcined, and fit to be reduced to powder in a marble mortar: Then take the other part, (being also all alive) and boil them in river ▽, like unto that wherein they were taken, then pour off the ▽, which being cold, put it in a wooden vessel, or of earth, and into about a pail full of this ▽, put about half a handful of the powder of the aforesaid calcined Cray-fishes, stir it well together with a stick, then let it stand to settle, without stirring it at all, and within a few days you shall see as it were many Atoms appear in the ▽, which are the breeding Cray-fishes, moving in the ▽ when you see them as big as a small button, you must feed them with bullocks blood, casting a little thereof into the ▽ from time to time, which in time will make them grow of their natural bigness. You must observe, that before you put the ▽ into the vessel, you must first put some sand into it, so much that the bottom of it may be covered about a fingers breadth.

To make Oil of Talc.

Take one part of Venice Talc, and two parts of pure Nitre, both in subtle powder, put them in a wind-furnace to calcine with a strong △ for seven or eight hours: Then take out the crucible, and beat the matter to subtle powder, and wash it perfectly well with fair ▽, till you have brought away all the saltiness of it; then dry the Talc well, and calcine it again with two parts of new Nitre, all as you did before, and dulcify the salt from it. Repeat these calcinations, and dulcifications four times, that the Talc may be perfectly white and well calcined, and in exceeding subtle powder: Then put it into a strong glass bottle, half full, and stop it close, and set it in a great quantity of ice or snow, that the extreme cold may penetrate into it (for therein consists the secret) but the ice or snow must not actually touch the bottle, but it must be set in a box of wicker, fit for it, made like a cage, that it be all open between the barrs or ofiers, and in two or three months all the Talc will be converted into a pure clear white liquor, which is excellent for the face and skin, and will make scarlet white, being dipped in it.

An Excellent Cosmetick prepared out of ☽

Take refined ☽, one part, *Sal gemmae*, two parts; beat the ☽ into very thin plates, stratifie them with the *Sal gemmae* in powder in a crucible well luted with another upon it: Cement for twenty four hours, then open the crucible, and if you find the ☽ well calcined, it is enough; if not, stratify it with fresh *Sal gemmae* and cement as before: Then edulcorate the ☽, with warm ▽; then grind it into a subtle powder, pour upon this powder a well rectified S. V.; digest until the S. V. is tincted blue; then decant, and put on fresh S. V. Repeat this until you have extracted all the tincture out

of the ☽. Then evaporate the S. V. with a very
gentle heat (or rather distill it gently) and there
will remain in the bottom of the cucurbite a matter
like *Pomatum*: Put upon this *Pomatum* a spirit of wine
rectified upon salt of Tartar, and after a little
digestion, distill off the S. V. in a retort, and
part of the tincture will come over with the S. V.
and will leave the *Pomatum* whiter than it was
before. Repeat this work with new S. V. (Tartarized)
so often, till the S. V. brings over no more
tincture, and that the *Pomatum* remains in the bottom
as white as snow, which is excellent to whiten the
face.

Another way to make Oil of Talc.

Reduce Talc of Venice into exceeding subtle
powder, mingle ℥ij. thereof with ℥ij. Of pure leaf
Silver, grind them exceedingly well together, to
incorporate them perfectly well: Then reverberate
them for fifteen or twenty days, after which grind
them again very subtly, and put them into a Matrass
of glass, set it *in fimo* for thirty or forty days,
changing the dung every six or eight days, that the
heat may be always in a good degree: You shall find
a good clear Oil, which will blanche pearls, the

face and skin in other parts, and do all the things that are said of Oil of Talc.

Another way to make Oil of Talc.

Take of Venice Talc in great pieces, as much as you please, make it red-hot in the △, then quench it by ceffing it into Oil of Tartar; fire it again, and extinguish it as before. Do thus two or three times, and it will be thoroughly calcined, so that you may crumble it into small powder with your fingers; beat it in a mortar, and pass it through a fine scarf of silk; what passed not, calcine it in a crucible, and extinguish again: It will be perfectly calcined by extinctions in fair ▽, but then it will require ten or twelve ignitions and extinctions.

Take your subtle Calx of Talc (which will be perfectly white) made either way, and put to it some distilled vinegar to swim two or three fingers breadth over it, and put it to digest in very gentle heat eight or ten days, and you shall see a beautiful Oil or Cream swimming at the top of the liquor, skim it off, and dry it with gentle △, and it will be a saline substance, which put into a bladder, and hang it in a well near the ▽ but not to touch it, and in a few days it will resolve into a pure Oil, which is excellent for the face. Or, with long remaining in a moist place, without putting it into distilled vinegar, this Calx will resolve into Oil: Try to extinguish the Talc in dew, and c. (Be sure that in all this work you touch nothing with iron.)

To burn holes in Glass.

When Mr. Gore would make a hole in the belly of a retort or Matrass, or receiver of glass, he did thus: Have some cotton-yarn well Sulphured, lay it round like a snake upon the glass, filling as much space as you would have the capacity of the hole, make a circle of tin, or the like, to keep it in (but be sure there be not the least moisture upon the glass, nor that it be very cold, for then it will break) set the cotton on fire with a burning coal laid upon it, and so let it burn on, putting up within its compass the burning yarn (with a Bodkin) if it chance to stretch, or swell wider than it should, make your heat and burning gentle and moderate at first, that you may increase it by degrees, by crumbling Sulphur in powder upon the burning matter, if you find it is needed.

When it has burned a while, try gently by touching it with a little stick of wood, whether the piece of glass under the burning cotton will fall in or out, but press not too hard, for fear of cracking that which should be whole; for when it is enough, it will fall in with the least touch, and leave a complete hole without any cracks in the glass besides. If you touch the heated glass with any moisture, you not only make that which you would have separated away to fall in, but you will crack and split what you would have remain solid. You may put a linen cloth in the glass, for the piece to fall upon, lest it should break the glass when it falls in.

A description of a most convenient and very useful furnace, which will not only serve for many operations, as Melting, Calcining, Vitrifying, Reverberating, Distilling, Subliming, Digesting, and C. But also for Coppelling, in small and great quantity, and that with the greatest facility that can be; so that neither Coals nor Ashes can fall into the Coppel, neither can the heat of the △ incommodate you by reflecting in your Face and Eyes.

The fabric of this furnace, with its structure, see in the next figure.

An Explanation of This Figure.

A.   Is the whole fabric, which may be built (of good bricks) about two foot four inches in length; one foot six inches in breadth; and about two feet and four inches high.

B.   Is the fire place, which must be round, of the best and hardest bricks (it may also be made of a fire stone) it must be eight inches deep, and eight inches diameter; at the bottom of it you may lay either a close grate, or a thick iron piece full of holes.

C.   Is the ash-hole and receptacle of the blast or wind issuing from the bellows, which must be very close, and the stopper fit the mouth of it exactly, to shut very close, so that the wind may find no vent anywhere out but upwards through the grate; this hole needs to be but four inches deep, from the grate down to the bottom; the ashes must be taken out of it from time to time, that they may not stuff up the place.

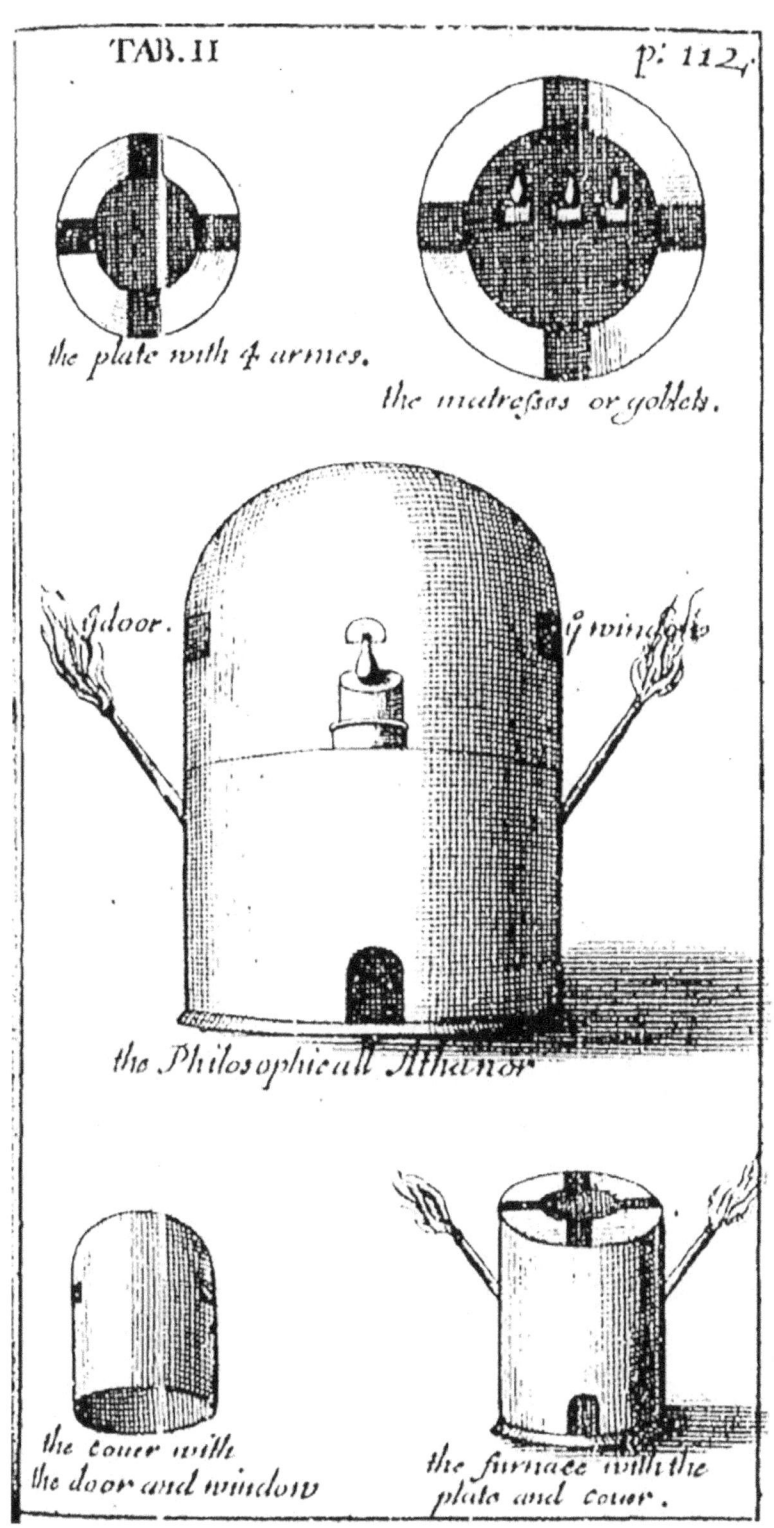

TAB. II   p: 112.

the plate with 4 armes.

the matresses or goblets.

i door.   ij windows.

the Philosophicall Athanor.

the couer with
the door and window

the furnace with the
plate and couer.

D.   Is the Coppelling-place, which may be about seven or eight inches in length, and six inches wide, and about four or five inches deep.

E.   Is the gap, through which the flame reverberates from the furnace into the Coppelling-place, it is about four inches wide, about two or three deep, and two inches in length.

F.   Is the Pipe or Chimney, which draws the flame.

G.   Is a hole, through which you may put fuel into the furnace, to avoid uncovering of it, when it is covered, as it must be when you Coppel; you must have a stopper exactly fitted to it.

H.   Is a hollow place under the Coppelling-place, which may serve to put Coals in.

I.   Is an Iron Hoop about an inch high, such as the refiners use to make Coppels in, you may have them of what bigness you please; at the bottom are fastened two flat Iron Bars, to hold the bone-ashes, having no other bottom.

To use this Furnace, you must have a pair of Smiths Bellows, of a midling size; which (if there be not room in your Laboratory to fix them below) you may fix above ground, and so they will not encumber your operatory below; and for the conducting of the Wind, you may have a square Pipe of Wood, made like an Organ-pipe, to come down from the nose of the Bellows through the Wall of the Furnace into the Ash-hole; and to the pearch you may tie a piece of Rope with a wooden handle at the end, to pull by in blowing the said Bellows.

In all the Operations I do in this Furnace (even when I Coppel) I use nothing but small Sinder from the Glass-House, which are not so heady as Charcoal.

Directions How to Coppel in This Furnace;

And First, How to Make a Coppel.

Take such an iron hoop, set it upon a sheet of brown paper to save the bone-ashes (that else you might scatter;) then fill it with a sufficient quantity of bone-ashes, first moistened with ∇ so that they hold together when you press them in your hand; stamp them well down with the end of an iron pestle, and make the Coppel very close and hard everywhere, then make it hollow in the middle, that it may hold the matter you mean to Coppel without running over; make it very smooth, then set it into

173

the coppelling place, and raise it with any sifted ashes, or with a piece of a fire-stone, so that the top of it may be level with the lower part of the gap. Then cover the coppelling-place with two bricks, (I use two bricks, because I can lift up one of them to put the metal into the Coppel, and leaving the other brick, the Coppel is not all uncovered, as it would be if it were covered with one whole stone.) Then having kindled the △ in the furnace, cover it as before, and blow the bellows, and the flame finding no way out, is forced and driven into the coppelling-place, where it reverberates upon the Coppel, which when you see that it is well heated, lift up one end of the hithermost brick, and put in the ♄, the quantity whereof must be proportioned according to the impurity of the matter you mean to Coppel; if it be Sterling ☽, you must take four parts of ♄ to one of ☽ if it be any other mixture of impure metal, you must take five, six, or seven parts of ♄ to one of the metal, according to the impurity of it. Govern the △ so, that the Coppel may always work and flow, and you may have a little space between the two bricks, through which you may look into the Coppel to see how it works, and if you see that it requires more flame than the Cinders, or Charcoal will afford, you may put into the furnace a round thick

174

piece of wood; but you must observe, that when the Coppelling-place is come to be red-hot all over; and the bricks also that cover it, the Coppel will then work with a very gentle △, so that then you must blow but gently; for if then you should give strong △, the matter in the Coppel would boil too fast, and would spatter about.

Instead of an iron hoop, many times I use but an Earthen Poringer, to make a Coppel, firing up the Coppelling-place with any ashes round about the Coppel, and find that it does altogether as well, only that it serves but for once. Note, that whilst you Coppel, or heat the Coppel, you may make Regulus, if you have occasion, or melt any other metal at the same time.

This furnace does far exceed any ordinary wind-furnace; for I can at any time make a parcel of Regulus, or melt any metal before the △ would kindle in an ordinary wind-furnace, and that with much less charge. In this furnace you may distill with a tilted retort in a naked △, by leaving two little holes in the wall of it to put two small iron bars in, to set the retort upon; you may also distill in it, in sand, both in a retort and in a cucurbite, by putting an iron pot into the furnace with sand, and laying some brick with clay about the pot, to enclose that part of the pot (or luted retort) that stands out of the furnace, and you may give what degrees of △ you will, from the first and lowest, to the fourth and highest degree.

THE END OF THE FIRST PART.

THE SECOND PART

Containing

Many Rare and Excellent and *Medicines*, Choice
Menstruums and *Alkahests*:
The true and only way to Volatize the fixt salt of
Tartar and to Corporifie *Spirit of Wine*, which is
Aqua Sicca in forma Salis
and is the true Vegetable *Menstruum*.

Never before Published.

LONDON
Printed for the Author. 1682.

A Real and True way

**to Volatilize the Salt
of Tartar, And Corporifie Spirit of Wine, as
it was wrought by a Noble Person beyond Sea,
and by him Communicated unto me.**

He took but lb j. of Tartar well Calcined, and dissolved it in the air, free from the sun, then filtered, and congealed in a glass vessel, then he calcined it as the *Dutch-man* does (in the next Process) to keep it glowing hot for six or eight hours, then dissolved it again in the air, and filtered, congealed, and calcined as before. He repeated this ten times, then dissolved it in distilled vinegar (he used *Spanish* and *French* vinegar;) the whole secret (as he said) consists in well distilling of the vinegar, which must be done in B. M. but it must be so gentle, that you may receive the flegm by itself; and as soon as you perceive that the drops come acid, change the Recipient, putting on another, and then distill the spirit with a stronger fire, so that you may tell eight or nine between every drop: Continue distilling until it become like a syrup, then change the Recipient again, and distill with a stronger fire until it begins to smell of the fire, and that it be almost dry. Rectify this last and strongest part by itself, and put it to the weaker part, (keeping the flegm by itself for another use) and rectify it together so often until there remain not the least spot at the bottom of the glass after the distillations, which must be to dryness every time, and every time in a clean retort: This is not a strong spirit of vinegar, nor need it be, but this will do the work. Then take $\frac{z}{3}$vij. or viij. of your salt of Tartar and dissolve it in as much of the said spirit, or more, as will dissolve it; let it stand, and it will settle some black feces; then filter it, and congeal, and calcine it as before, but not with so strong a fire; it must be scarce red-hot, and one hour will suffice; grind it while it is hot, and dissolve it again in new distilled vinegar as before, let the feces settle, then filter it, congeal and calcine again as before. Repeat this, till it leaves no feces behind, which will be in seven or eight times, if you have wrought well; then when it is very dry, take $\frac{z}{3}$j. of it to make a trial, put it into a clean glass body, and pour upon it so much highly rectified spirit of wine, as will not only moisten it, but that it be very thin; let it stand thus twenty four hours close stopped in a very gentle heat, that it may be but blood-warm; then distill with a gentle fire. If the spirit stays, and the flegm come away, then proceed with the whole parcel; but if not, you must continue the

179

dissolution in distilled vinegar, filtration, coagulation, and calcination, as before, until you find (by trying) that the spirit stays with the salt, which it will do in a few times: Then proceed with the rest of the parcel in the same manner as you did with the ounce; continue the imbibition and distillation with spirit of wine so often, till the spirit of wine come away as strong as it was put on. Then here lies the secret, to sublime it: Dissolve the said salt of Tartar impregnated in the flegm of your distilled vinegar, or in a very weak spirit of wine, using no more of the dissolvent than will dissolve it; shake them well together, and it will instantly dissolve all the best and finest part of the salt of Tartar, and leave the course part of it, for that will not so easily dissolve. Pour off the dissolution, and filter it, then put it into a cucurbite, and distill off the flegm off the vinegar, or the weak spirit of wine, and then will the dry spirit, or Aqua sicca ascend like the purest icicles dry that ever you saw; and this is the true volatile salt of Tartar, and spirit of wine, in forma Salis, and is the vegetable Menstruum, which will dissolve leaf ☉ into an Oilish substance in a very gentle heat.

The Tartar which remains in the bottom of this sublimation you must put to that which was left undissolved by the flegm of vinegar, or weak spirit of wine, and proceed to fix more spirit of wine upon that, being first calcined, but not so long, nor with so strong a fire as formerly; and then dissolve it once in the air, and it will leave more feces at this time than at any time before; then filter and congeal, and dissolve it in distilled vinegar as before. And now you shall do more in three times than you did all the times before, for the Tartar is altered in its nature; then proceed with it as before, imbibing with S. V. And thus you may fix as much S. V. as you please, and sublime as many pure and clear crystals as you please.

Note, That when your spirit of wine is fixed on the Tartar, it will be as sweet as sugar; but when it is separated, as above-said, the Tartar will be of its old Nature, but fit to be impregnated again with much less trouble.

The Dutch-mans Process

Of Volatilizing Salt of Tartar, and Corporifying Spirit of Wine, is thus:

He dissolves his Tartar calcined in distilled rain ∇, and being settled, filters and congeales, then calcines it for six or eight hours, so that it is only glowing hot, and no more: Then powder it, and dissolve it, filter, congeal, and reverberate, as before; and this he does sixteen or eighteen times, until the Tartar settles little or no feces. Then take four or eight ounces of it (or what you please) and put it into a cucurbite, and pour upon it the best rectified spirit of wine, so much, that it may be well moistened, but not to swim over it. Then he digested it in B. M. for a day, and then gently distills it off, but the spirit of wine ascends, and will not stay; when it is dry, he puts the spirit of wine back again upon it, and distills as before; and this he does so often (twelve or fourteen times) and then it will hold the spirit of wine, and the flegm will come away: This he does with new spirit of wine; and when he finds that a pretty quantity is congealed with the Tartar, he grinds it (being very dry) and mixes it with three parts of good Bole-Armony, and puts it into a retort, and distills a spirit from it by degrees of fire, forcing it strongly at last, and it will be a yellow spirit, which he deflegmes once, and then dissolves his Calx of ☉ in this Menstruum, which by digestion extracts all the tincture of ☉, and leaves the body white, tinging itself of a pure red. Of this he gives two or three drops in a little sack, which does miracles (as he said).

He made his red Calx of ☉ thus:

Dissolve fine ☉ past through Antimony in A. R. then put water into it, and then cast in ☿, and all will become like a *Hepar*, then wash the *àààà* and grind it with three times as much prepared salt, and distill it in a retort; then edulcorate the Calx, and grind ℥j. of it with three of Cinaber, and then reverberate it by degrees gently: This he does twice a day, and repeats it eleven or twelve times, and it will be a most subtle red Calx, like Scarlet in colour.

Elixir ex vino & Sole.

Take the best odoriferous Rhenish-wine five gallons, separate its spirit and salt, rectify the one, and purify the other; then acuate the spirit with another prepared salt, and at last join it to its own pure crystalline salt; then is it a true Aqua vitae Philosophorum. This must dissolve a well prepared Calx of ☉, and by a continued circulation unite with it; then by sublimation be fixt together, and lastly, by solution and coagulation become incombustible Oil, which is a great medicine.

Monsieur Toysonnier Wrought Thus:

Take fresh Urine of young boys, fill one pot with it, and evaporate it away, next morning put on fresh, and evaporate; do thus three or four days, then evaporate to a honey, and that you feel a ponticifie smoke from it, then cease, and put your honey into an Earthen vessel, and expose it to celifie in the air. As soon as it is cold, it will be hard, but the air will resolve it: Make thus what quantity of honey you please: Celifie them four days, then have another Earthen pot, covered with a reversed one that has a hole in the bottom, fasten thereon a neck of a cucurbite of glass, ten or twelve inches long, upon which a retort, with the bottom out for a head, to which fasten a great ballon. He did put fifteen pounds of honey into his pot, and with a gentle fire first distilled off the spirit and volatile salt; these he put upon new honey, and in balneo distilled a purer spirit and volatile salt; (the flegm that followed, if put upon new honey, will become pure spirit and salt) draw the fixed salt out of all the Caput Mortuum; put ℥xij. of the spirit upon as much pure spirit of wine, and it will coagulate it all into a perfect dry salt: Mingle these ℥xxiv. of volatile salt with ℥vj. of salt of wine, ℥iij. of volatile salt of urine, and ℥iv. Of ☿ precipitate, and put them into a body with head, limbeck, and receiver, and sublime with gentle heat: Part comes over in spirit, and part rises in salt. Take ℥xiv. of salt, and vij. of spirit, and ℥ß. of Calx of ☽ and distill with exceeding gentle heat in a body and head with a receiver, a liquid spirit will come over and a white salt sublime into the head: Put all back upon the cake of ☽, and distill as before. He has now repeated this work eleven times; at the first, the spirit and salt were ten days rising from the ☽, but afterwards seven or eight: The junctures were all perfectly shut, yet above half of the volatile matter was vanished. The salt of wine was made thus: (Spanish wine gave none, but French did pretty store.) After you have drawn off the spirit and the flegm, evaporate the residue (very gently) *usque ad pelliculum* then set in a cold place, and in fifteen days there were many crystals in it; wash these with the flegm of the wine, from the blackness and foulness that is upon them. The ☿ precipitate was made thus: Dissolve ℥iv. Of ☿ in ℥x. of A. F. made of two parts of vitriol, and one of Nitre. Extend the solution, by pouring a great quantity (eight or ten pints) of fair ▽ upon it; then pour upon it a *Lixivium* made of the fixed salt of wine and fair ▽.

He made his *Lixivium* of $\text{z}$iv. of fixed salt, and but one of the precipitate; wherefore he poured upon the liquor that he poured off from the ☿ precipitate about half a pint of the spirit of wine, and then the ☿ precipitated all down. Take both the precipitates, and wash them a little from the spirits of the A.F.

(Hartman) This relation is of Sir K. It was done by his operator Monsieur Toysonnier, in his operatory in the Piazza in Covent-Garden.

The Menstruum Coelicum Exuberatum,

to dissolve ☉, and all Metals, and carry them over the Helm. Wrought by Dr. Clodius, And by him Communicated unto me.

Put lbj. of perfectly rectified spirit of urine to lbiij. of a perfectly rectified spirit of wine, and it will coagulate it all to a drop into a firm salt; sublime and distil this, and about lb ß. will sublime up in a most pure active dry salt, and about two spoonfuls will come over in a fiery liquor, and the rest will be a stinking flat flegm remaining in the cucurbite. Add spirit of wine to your two spoonfuls of liquor, so much as to make it up lbiij. put this to your lb ß. of dry sublimed salt, and all

186

will be a Coagulum, which distill and sublime as before, and you shall have about two spoonfuls and a half of fiery liquor, and about $\text{z}$ij. more than before of sublimed salt, and in the bottom will remain a stinking flegm as before. He repeated this twelve times, still with fresh spirit of wine, and every time the quantity of the fiery spirit increases (drawing still some little addition of dry salt from the spirit of wine) till at length all the dry salt comes over in liquid fiery spirit, which he calls *Menstruum Calicum*; then it will dissolve ☉, and all metals, and carry them over the helm with it. This poured upon salt of Tartar, will presently dissolve it, and carry it over the helm with it: You may also multiply it as much as you will with pure salt or spirit of urine.

To add ♀ to it, do thus. Take ☿ well purified (if Spanish) shaking and washing well with spirit of wine or distilled vinegar, till it yields no more blackness, will serve; but if other, sublime it sometimes to perfect purity. Then he put lb v. of it into a retort, joining a large stone receiver, and gave strong sudden △. Repeat this till you have Mercurial ▽ enough, which will be quite insipid: Put this ▽ upon purified running ☿, and digest them together thirty days, and all will be a viscous Mucilaginous Matter: Distill off the flegm, and you

187

shall have an Oil remaining; put upon this Oil your fiery spirit of urine, and spirit of wine (twelve times repeated, as is mentioned) and the spirit will resolve the Oil; then distill them over together, and you have the *Menstruum Calicum Exuberatum*. To *âââte* Regulus of ♂ with ☿, you must pulverize the Regulus grossly, as Bay-salt; then strew it gently upon the ☿ in a Matrass, and digest three or four days; then grind together, and it will *âââte*, Thus you may do with Reg. and ☽. He found difficulty in doing it with and Reg. but after digestion, he let it stand a little while with ▽ and salt upon it, and it *âââted* of itself with the ☿. If you digest eight days, the blackness you wash away is a combustible sulphur, like powder of coal. He made Mercurial ▽ thus: Put store of quick dry sand upon ☿ in a cucurbite, and distill it in very gentle fire, and most will come over in ▽: Put this ▽ upon new ☿ and digest, and it will bring it into an oily substance; mingle this with your Coagulum of spirit of wine, and spirit of urine (See Lully's Eight (8) Experiments) and make an Alkahest of these, which will be perfect when it is *âââte* with ☉. Then proceed as Lully teaches.

 The said Dr. Clodius told me also, that the great secret of purifying all salts and vitriol,

etc., consists in the purifying the Menatruum (i.e.
▽) for if the ▽ have feces, in which you dissolve
thevt it rather increases their foulness. He does
thus:

Set Equinox rain ▽ (pure dew is better) to
putrefy in glass vessels, slightly covered (only to
keep things from falling in) in a cellar; in six
weeks the putrefaction will be finished, and all the
feces fallen to the bottom; filter it, set the clear
to putrefy again, which will require longer time
than before:

If you put some quick dry sand in it, it will
help the putrefaction much the sooner, drawing down
the foulness to it. In this purified ▽ dissolve
your salt, vitriol, salt of urine, etc.; And you
must have a gallon of this water to ℥j. of salt; for
such dilating of the salt made the Menstruum lighter
than the feces, and therefore they fall down. Mark
the end of Isaac Hollands Process upon Vitriol,
where he directs us to dissolve it in fair ▽
distilled off then filter and congeal, and this you
may repeat two or three times: Then take nine parts
of this pure salt, and put to them one part of pure
spirit of wine, digest them together for seven or
eight days, then distill off the liquor very gently,
and about fourteen or fifteen parts will come off in
insipid flegm, and only one part, or a little more

will remain with the salt in a dry substance. Repeat this nine times with fresh spirit of wine, till you have employed as much spirit of wine as you wrought upon salt, which will be increased scarce one part: Put it then to sublime, and every whit of it will rise in a pure sublimate, excepting a small parcel of feces that will remain in the bottom: You may multiply this sublimate as much as you will, by addition of spirit of wine to it; then distilling away the liquor, for the ✳ of the spirit of wine will adhere to the pure salt in a pure saline form, and the rest will come away in an insipid flegin. But in the end, this salt will be apt to come over in an oily liquor, and sometimes after that, in form of salt again: But he cannot yet penetrate into the causes of these bodies coming over sometimes liquid, sometimes dry. Incorporate some of this salt with pure spirit of wine, and it will dissolve ☉, and all metals.

To Prepare A Most Excellent Medicine

with this Mercurial Water proceed thus:

Take of the afore-said Mercurial ▽, and of spirit of wine distilled three times upon honey, and then rectified upon salt of Tartar, of each equal parts; distil them together, until they be well united: Then to six parts of this Menstruum put one part of a spongy Calx of ☉, digest them together, until the ☉ is totally dissolved, except a little white earth, which will remain in the bottom. Then distill in a retort in sand, and cohobate so often, until the ☉ come over into the receiver. Then separate it, by distilling in a cucurbite, and there will remain a red Oil like a Ruby in the bottom; whereof one drop in some fit vehicle, is admirable for health. This dissolution of ☉ has a most Odiferous scent, better than Amber and Musk.

The Lunary ▽ of Paradise,

or the Celestial Eagle of the Lunar Sphere, which is Lully's true Spiritual Lunary.

Dissolve ☽ in A. F. and precipitate it with spirit of salt, dry the Calx, and mix it with equal parts of Calx of ♃ and thence distill the volatile crystals or butter s. a. Expose this butter to the air to resolve into liquor, that by that means it may make a dentifique attraction of the spirit of the world, which specifies itself by this magnet. Put the clean liquor into a cucurbite, and digest with a Lamp △ for fifteen days, at the end of which time there will distil over an AEtherial liquor, which is the Lunary; with which you may work Miracles in Physick, especially in all diseases of the head and brain; the dose is the same with the former, mixing it with a fit Vehicle; Note that it turns all liquors into milk, and is sometimes Emetick. For transmutation, deflegm this ▽ as before and you shall have the metalline gluten; which being digested per as, will become a white powder of projection; and afterwards a red one: But it will be better to add a tenth part of leaf ☉ or of the Sulphur of ☉ made by the Sal Enixe. It is multiplied by new addition of the Lunary or Gluten. Note, that this liquor is the Sommet of the Lunary: For the volatile crystals of ☽ are the simple Lunary; but this coelestial ▽ is the spiritual Lunary, or the highest point of the metalline salts:

Note, that you may draw the red and white Oil of what remains, and proceed as before.

Water of Paradise of Saturn,

or Jupiters Celestial Eagle.

Dissolve ♄ in common A. F. and it will precipitate into a potential ☿: dry it, and mix it in great quantity with equal parts of Calx of ♃ or with ♂:

192

Draw the volatile crystals or butter from this, which resolve in the air: Put the liquor into a glass cucurbite, with its head and Recipient, digest with a gentle lamp △ the space of fifteen days; nothing will come over by distillation, but only the liquor will maturate, and become red, and afterwards within twenty, thirty, or forty days the Idea of ♄ will rise invisibly, and distill into the Recipient, which is the ▽ of paradise.

This ▽ Cures All Saturnian Maladies,

and Melancholy, being mixt with S. V. You may give it in all Inflammations both Inward and Outward: The Dose is Equal with the Former.

To use it for metals, put this ▽ of paradise into a small cucurbite and deflegm it with a Lamp △ and there will remain the Saturnean Gluten, the Philosophers Gum, or ☿ of the wise, A Lot, &c. Digest *per se*, or add a tenth part of ☉. This gum is the true metalline radical moisture.

About Vitrum Antimonii, and its Tincture.

Monsieur Borel told me, that he had observed this in making the tincture of ♂ (by distilled vinegar upon the glass of ♂) that when he went to dulcify the salt that remains in the tincture after the distilled vinegar is evaporated away (as Bas. Val. teaches) he could never perceive that the ▽ evaporating carried away the remaining salt of the vinegar, but still when the ▽ was gone, and left the powder dry, it was as salt as ever before, and was of a blown gray colour. But this he observed, that after four or five times dissolving in ▽ and evaporating, the tincture precipitated down very red, and the salt of the distilled vinegar remained dissolved in the ▽, so that he then poured off the ▽, and dryed the powder, which then was exceeding red, and perfectly dulcified: But after thus severing of the salt from it, S. V. would not touch upon it, and extract it any further: Peradventure a Tartarized S.V. will do it.

He also told me, that in making the *Vitrum Antimoni* for this work, the mystery to have it certain and constant, consists in this; That after you have calcined your by ♂ long and gentle evaporation and stirring, so that it smokes no more;

194

and when you have put it in the crucible to melt into glass, you must put to it a little piece of a coal to burn with the ☿, and set the sulphur of it on fire, which will make a little Regulus fall down to the bottom, and the glass will be pure clear and red, whereas if no coal fall or be put in, it will be black and muddy: And that which made this work of vitrifying the ☿ prove uncertain, is, that sometimes some coals fall into the crucible (as one gives great heat) without the artist taking notice, and then the work proves well, but if no wooden coals fall in, the glass proves not as it should.

Monsieur le Fevre told me, that when he makes the tincture of Vitrum Antimonii, he observes, as Monsieur Borel said, that if the extract of it made by distilled vinegar is perfectly dulcified from all saltiness, the S. V. will not touch upon it; and if it is put upon it in dry powder, containing the salt of the distilled vinegar with the tincture of the ☿ , it will draw with the tincture some nocuous spirits from the salt; therefore he does thus: Evaporate fair ▽ from it two or three times, the last time leave it very moist; then put S. V. upon it, and it will presently impregnate itself with the tincture.

A white Spirit of Sulphur to dissolve ☽ and ☿;

given me by Monsieur Bugneau.

Take the black spirit of Sulphur, made *per Campanam*, put it into a glass retort well luted (for fear of breaking) all but a patch at the top as big as a Crown-piece, that you may thereby see (holding a candle near it) in what state the liquor, therein contained, is: Distill in sand, till all the flegm is come over, and that it begin to drop very sharp, which happens to him after ℥vj. or vij. are come over, of lb j. of spirit put at first into the retort. By this time you shall see a little ✳ sublime up to the discovered place of the retort, and a brown circle of earthy substance swimming upon the liquor about the sides of it joining to the glass: You must now give the △ so quick, that the liquor boils a little, and presently you shall see it turn all white, and the brown Corona of earth becomes white: Then let the fire die, and when the retort is cold, pour out the spirit, which will look like rock ▽ and will leave some dregs behind; you will have about ℥ix. of this spirit, put it upon ℥ iv. of ☽ in leaf, and distill it gently off, and after a while you shall see your ☽ quite dissolved into clear liquor; let it cool, and the ☽ will become a cake of crystal, and some liquor will swim over it; pour off this liquor, and put ℥ij. of fresh ☽ to (which now may be in small grenaille) and it will dissolve this as the former, and become a crystal by cold: The liquor that you then pour off will dissolve ℥ij. more of fresh ☽, doing as before. Now the liquor that remains after this third solution of ☽, will dissolve ℥ij. of running ☿ into a crystalline substance as the former.

This spirit of Sulphur thus rectified, being used inwardly (before it is used with ☽) is much stronger than when it is black, at the first drawing, and is much more grateful to the taste, being mingled with ▽ or other vehicle.

A Universal Medicine, from ☉ and ♂, and c.

Take of the ☿ prepared, as shall be taught hereafter, ℥j. of the tincture of ☉, afterwards set down, ℥j. mix them well together in a glass mortar, then put them into a small Matrass, and digest them with a Lamp △ with one wick only for ten days; then digest for ten days more with two wicks, then with three, and lastly, with four wicks, which makes

197

forty days digestion in all, at the end of which you shall have a red powder as red as a Ruby.

This powder is a Universal Medicine for the greatest and chronic diseases: It cures the gout, dropsie, palsie, French-pox, plague, leprosie, the evil, small-pox, and measles. Its visible operation is by stools, by urine, and by sweat: The dose is from gr. iij. to iv. or v. in conserve of borrage or violets.

To prepare the ☿ for this Work.

Take gravelled ashes, (or instead thereof you may take the ashes of dryed and burnt Lees of Wine) and of Quick-lime, of each equal parts, boil them together in ▽, and make a Lixivium, which filter. Take ℥iij. or iv. of ☿ vitae, put it into a Matrass, and pour upon it of the aforesaid Lixivium, so much as may cover it the breadth of four fingers; digest with the second degree of heat, for three or four days, the Lixivium will extract the tincture of the ☿ vitae; then decant, and put on fresh Lixivium, and digest. Repeat this, till you have extracted all the tincture of your ☿ vitae, and the powder be well attenuated: Then mix this powder with equal weight of sublimed ✳, incorporate them well together with double as much of Oil of Tartar, then set it to putrefy *in fimo* for thirty days, changing the dung every sixth or seventh day. Then put your matter in a marble mortar, and grind it well, adding a little warm ▽ to it; then add a little more water, but a little hotter than the first, and grind it well; then let it settle, and decant the ▽ and put on fresh warm ▽, and grind as before, then let it settle, and decant the water; and put vinegar upon it instead of ▽ and grind it, and you will see in a short time the powder converted into running ☿. Note, that if you sublime Regulus of ♂ with four times as much ✳, it will sublime with it in very red flowers; out of which in the same manner you may extract ☿.

To prepare the Tincture of ☉ for this Work.

Take fine ☉ in thin plates, dissolve it in A. R. then pour into the dissolution some ☿, and a fourth part of A. F. keep it in digestion until the ☿ is all dissolved; the dissolution of ☉, which was of an orange colour before, will now be white and clear, and the ☉ will precipitate to the bottom in a very subtle and spongy Calx; decant the clear, and edulcorate the powder of ☉, till it be freed from all Acrimony, then dry it.

Then take fine Pumice stone, and make it red-hot in a crucible, then extinguish it in vinegar; reiterate the ignitions and extinctions five or six times, then reduce it into subtle powder, which ignifie again for half a quarter of an hour, then make it as subtle as you can. Then put a bed of this powder into a crucible, about a fingers breadth, upon that put a bed of your powder of ☉: Continue thus stratifying until all the powder of ☉ is in, then cover the crucible with another, and lute them well together, and put it into a glass oven where they prepare their matter, so that the crucible may be always red during twenty four hours, and that the matter in the crucible may not melt. Then take out the matter out of the crucible, and pulverize it; then put this powder in a Matrass, and pour upon it of the following dissolvent, so much as may cover it three fingers breadth, digest it in ashes for thirty four days, within a few hours you will see the dissolution tincted of an orange colour; after four days digestion decant the tincture, and pour on more of the dissolvent, digest as before. Repeat this till you have extracted all the tincture of your powder; then filter all your extracts, and evaporate with a gentle △ to dryness, and you will have a yellow powder of an orange-colour; put this powder into a Matrass, and pour upon it a S. V. prepared as shall be taught hereafter; digest it, and in two days the S. V. will be tincted as red as blood, which decant, and put on fresh S. V. digest and decant. Repeat this so often till you have extracted all the tincture out of the powder: Then distill off the tincted S. V. in B. with a gentle heat to dryness; and thus is the tincture of ☉ prepared for this work, to be used with the said ☿ of ♂, as is said above.

Note, That if you digest and circulate this tincture *in fimo* before you distill the S. V. from it, and then distill and cohobate two or three times, and abstracting half the S. V. from it, you will have a kind of an Aurum Potabile, which is a very great corroborant in the greatest weakness: The dose is five or six drops in any convenient vehicle.

The Dissolvent.

Melt salt in a crucible, then take lb j. thereof and pulverize it; mix this powder with lb iij. of honey, boil them together in an iron kettle to the consistence of a suppository; then cast this matter upon a smooth stone, and being cold, pulverize it, and put it into a retort; pour upon it distilled vinegar rectified, lb iij. Digest for twenty four hours, then distill in sand by graduated △, giving strong △ at last for six hours, that the retort may be red; then let it stand to cool the space of twelve hours: Then distill this vinegar in a cucurbite in ashes, separating the flegm, rectify it three or four times more, and it will be white and clear; before it was yellow.

To prepare the Spirit of Wine,

fit for this Tincture of ☉.

Take salt of Tartar well purified by several dissolutions, filtrations, and coagulations, and then reduced to powder, ℥iv. which put into a re-tort, and pour upon it lb ij. of rectified spirit of wine, let it stand so twenty four hours, then distill only lb j. of it in ashes, and you shall have an excellent spirit of wine, fit to draw tinctures. In the same manner you may extract the tincture of Coral, putting the Corals whole with the Pumice-stone, which by its dryness will extract the tincture of the Corals, leaving them as white as starch.

In the same manner you may also extract the tincture of ☽, which will be blue.

(Sir Kenelm D.) This process was given to Monsieur Vrto, Physician of Burges, by Monsieur Mayo, Sieur de Vancours. This Monsieur Mayo was a great friend and confident of Monsieur de la Violette, who gave him this operation, and they made it together. He said, that this was the solidest and best thing that Monsieur de la Violette had. He gave this to Monsieur Vrto in acknowledgment of a very great good turn he had done him, and after Monsieur Vrto had refused to receive of him a present of great value.

A great Corroborant and Sudorifick,

wrought by Monsieur Du Closs, Physician at Paris; given me by him the 16th of August, 1660.

Dissolve ⊙ by means of Salt, Nitre, and Allom, &c. after Zwelfer's manner; then evaporate away the ▽, and put S. V. upon the remaining powder, and it will go all into a tincture, or rather all the ⊙ will dissolve in the S. V. leaving the salts, most of which will precipitate in the S. V. Then he precipitates the ⊙ with Oil of Tartar, and washes and dries it, then reverberates it, and it is in a deep red powder; and this he called Crocus Solis: (But it is not so, almost all the ⊙ remains still in the solution (which is yellow) and Oil of Tartar will not precipitate it, so that it is rather the salts that remained in the S. V. and a little mingled with them): But take spirit of honey (the vinegary spirit) two parts, and one part of S. V. and pour this upon the solution, and all the ☉ will precipitate like a green mud; pour off the liquor, and put fair ▽ to the precipitation, and some ☿, and so you may have all your ☉, which when it is dry, will be a deep-red powder, but if you reverberate it, and *áááte* it with ☿, and grind it with Sulphur, and then burn and reverberate it, it will all fly away: And this is his best way of calcining and opening ☉.

Upon this Calx of ☉ he put his Menstruum, and in twenty four hours it will tinct itself as red as blood, which if you digest long, an Oil will swim upon it; he evaporated the Menstruum till it was thick, and digested that with a Lamp furnace.

His Menstruum is thus made:

Take pure S. V. and pure spirit of urine, *ana*, put them together, and distill off the S. V. with very gentle heat, there will remains a flegmatick liquor in the bottom: Cohobate the S. V. upon it till there remains only perfect flegm in the bottom, and that all the spirits and volatile salt of the urine be in the S. V. This is a great dissolvent and alkahest; but it will be stronger if you work it again with new spirit of urine, and so you may make it as strong as you will: But this does not have the properties of Helmont's pretended alkahest, to come off from the body it has dissolved, as strong as you put it on, for it leaves much of the saline spirits with the opened body, if you distill it off: He found some running ☿ in the filters after he had dissolved the ☉ only as far as Zwelfer teaches; which solution opens it exceedingly, and renders it apt to mercurialization; but he uses most the following Calx of ☉: Make an *ááá* of ☉ and ☿ in due manner, which grind well with flowers of Sulphur, and set it upon coals, and so make a Calx of ☉ (*ut artis est*). Repeat this calcination two or three times, then take the Calx of ☉, and grind it exceeding well with twice as much pure decrepitated salt; put these into a crucible, which cover well, and set it to cement or reverberate during six hours (or more) in a furnace where the heat may be increased by degrees, so that in due time the crucibles become red. Continue so a pretty time, but have a care the salt melt not: When it is cold, take out the matter, and grind it well, and pour hot ▽

upon it, to dissolve all the salt, and filter it off, and pour on more ▽, doing so till you have severed all the salt from the ☉ (as also a white earthy substance, that will swim upon the ▽) then dry the ☉, which grind again with double its quantity of prepared salt, (the same salt will serve again when the ▽ is distilled from it) and cement it, and work all as before, taking care always, that the ▽ settle well to the bottom after you have stirred it in the ▽. Repeat this six, seven, or eight times (the more the better) till the ☉ come to be all a gray or white powder: Then cement it with double its quantity of pure salt of Tartar, in the same manner as you did with salt, and do always all as before. Repeat this two, three, or four times, dulcifying it every time very well from the salt: Then put upon it (being very dry) the Menstruum of S.V. and Spirit of Urine, mentioned before, and it will be tincted blood-red in twenty four hours: Pour off that, and put on more, till you have drawn out all the tincture, which distill in a cucurbite with very gentle △, till it becomes a gum, of which he put ℥j. into a pint of sack, and give a spoonful for a dose. It is a mighty corroborant, as also a sudorific, where nature requires it. It will make one sweat twenty four hours.

The manner of making his Menstruum, is, to put the two spirits into a long cucurbite with a narrow mouth, on which he put a head, fitting it in the orifice, but very large in the body of it, and so distills off his S. V. and cohobates it upon the same spirit of urine, till the volatile salt is drawn out of it, or upon new, as you see on occasion.

Quære, of putting this Menstruum upon a spongy gray Calx of ☉, made after VanDykes way.

The Metalline Aureal ▽,

or the Aethereal Aurum Potabile, which is a very great Medicine for the Gout: it is the true Hermaphroditick Bath.

Dissolve ☽ in A. F. then precipitate it with spirit of salt, then edulcorate the powder and dry it, then mix it with its weight of ♂ (or Calx of ♃) distill a transparent butter thereof: Take of this butter one part, mix it with as much of Calx of ☉ (made by dissolving ☉ in spirit of salt) digest them together, until they are reduced into a liquor: Distill this liquor in a retort, the spirit of salt will come over first, and then will follow a red butter, which is the great *Chalybs*, which resolves into a liquor in the air; put this liquor into a cucurbite, join a head and receiver to it, and then digest with a Lamp △ for fifteen days, then an Aethereal liquor will begin to come over in an invisible form, which will distill into the Recipient: Deflegm this liquor until you come to the Eagles Metalline Gluten; which is digested (either *per se* or with ☉) into a true physical stone: When it is in an Aethereal liquor you may take two drops of it in some cordial spirit.

The Eagles Gluten, or ☿ of the Wise,

or Metalline Menstruum; with which and Lions Blood is made the Metalline Stone.

The gluten is of divers sorts: The first is altogether mineral, and is drawn from ☿ and ♂: If you join Sulphur of ♄ with this gluten, you may make a medicinal stone of it. The second is metallick, viz. Saturneal, Lunary, and Aureal. The third is partly mineral, and partly metalline; as for example, when one draws a liquor (which does not wet) from ☿ of ♄ (that is to say, from its repercuted Calx) and ♂, which is the magnet of the

208

spirit of the world; then draw the gluten as you know. The gluten is mineral and metalline, and is sufficient to make the physical stone of it, both mineral and metalline.

Note, that if you digest *per se*, what sort of gluten soever, you may make the physical stone of it. But for to shorten the work, you may add ☉; for all metalline or mineral gluten contains in itself its internal Sulphur, which may be coagulated and fixed into a true Aetherial *Panacea*. But it is better to add this solary ferment, as shall be said hereafter. Wonderful things may be performed (both in Physick, and in Transmutation of Metals) with any sort of gluten, either mineral or metalline. The ▽ of paradise differs not from the gluten, except that it contains some parts more liquid, and as yet full of flegm, as shall be shown.

Water of Paradise,

or of the Hermetick Eagle, whereof are made unheard-of Medicines, and Powders of Projection.

The ▽ of Paradise is a certain fiery of Aethereal ▽ drawn from Coelestial bodies, chiefly from ☉ and ☽, without the mixture of any waterish flegm; so that, what is attracted is the universal spirit, the informing form of the elements, that of the world, influence of the stars, soul of the world, the vital nutriment, latent in the air. This ▽ is most potent to drive out all diseases, it being altogether astral, and need not be taken by drams, scruples, or grains, but the twentieth part of a grain is sufficient for a dose; yea, almost the vapour only of this gluten suffices, as you shall see: it is attracted by several things, or (to speak plain) there are several things which attract it from the stars; first, by Sendivogius his magnet, or *Chalybs*, but it requires a longer time to have this gluten, or this Philosophical ▽, which is all, to all universal; for it requires seven months to prepare this universal Menstruum, after you have the salt of nature; which is a thing indeterminate, and requires a metallick ferment, specifically, to specify and make it determinate. This most noble way is clearly and neatly shown by the author: But there are other ways, which are shorter, by which this spirit of the world is attracted by several magnets, whereof shall be spoken hereafter. Note, that the physical stone may be made of all sorts of waters of paradise; for it is the Philosophical ☿ which is sufficient for himself and for thee; for it contains in itself a pure Sulphur, which may be congealed into a *Panacea*: But for to shorten the work, the solar or lunar ferment is added, to the end that this gluten, or fiery ▽ may be sooner congealed and fixed: So that, besides this *Generalissima* way, or this Universal Stone of the Philosophers, there are five other stones; to wit, first, the simple mineral, made of ☿ *per se*, or with ☿ and ♂ with the Sulphur of ♄. The second is the simple metalline stone, made with ☽ only, with ♃, or with ☉ and solar ferment. Thirdly, there is a stone which is partly metallick, and partly mineral, made of ♄, ☿ and ☉, whereof Artefius, Flamel, Pontanus, Zaichair, and others have written. Fourthly, there is a vegetable stone. Fifthly, the animal stone. We shall treat of all these stones, under the names of the ▽ of Paradise, or the Hermetick Eagle, or Virgins Milk.

Water of Paradise of Common ☿,

or Hermes his Eagle, of the Terrestrial And Celestial ☿.

Sublime ☿ three or four times with Salt, Nitre, and Vitriol; then dissolve it in A. F. and digest, then by distillations and cohobations, unite the Salts Armoniac of the A. F. to which (to have it more powerful) you may put an eighth part of ✳:

Distill and cohobate so often, till the ☿ comes to be like wax, and that it dissolves easily *in humido*. Then dissolve this matter *per deliquium*, that it may attract the ▽ which is contained in the air: Put this liquor in to a small cucurbite, join its head and Recipient, and digest with very gentle △ with a Lamp. Nothing will come over during fifteen days, but afterwards, there will come over an Aethereal liquor, which is the ▽ of Paradise: Two drops of this ▽ put into ℥iv. of S. V. is an excellent medicine against the pox, for it is the planet ☿. The dose is one spoonful. The physical stone is made of this Virginal, or Astral Milk, to wit, distill its flegm, in a small cucurbite, with the same Lamp △, and the gluten or mineral gum will remain in the bottom; of which by digestion is made the physical or medicinal stone. But note, that if you add ☉, the operation is sooner accomplished. Note also, that if you cast one drop of this ▽ of Paradise upon a thin plate of ♀, or of ♂ it will penetrate and whiten it through and through, before it is fermented with ☉. Note also, that that which remains after the distillation, will serve also. If you would then make a stone, different from that which is made with the Virginal Milk only, proceed thus: After you have distilled the ▽ of Paradise, distill over with a gentle △ in ashes what remains, and you shall have a white Oil; then force over the remaining part in a retort, and you shall have a red Oil; cast away the remaining feces. Take one part of the red Oil, and four parts of the white Oil, and eight parts of the ▽ of Paradise, put them into a Matrass, and digest them in an athanor until all the colours appear one after another, and that the gluten be fixed into white. If then you augment the △, it will become a red medicine, of which you may make projection thus: Take an hundred parts of ☿, heat it in a crucible, and cast upon it one part of this fixt medicine, and all will be a medicine; whereof cast one part upon another hundred parts of ☿, stirring it with a stick; then melt them together. Cast one part of this medicine upon an hundred parts of ☿ and all will be converted into ☽ or ☉, according to the tincture. In this manner, all metals and minerals may be reduced into tinctures by their ▽ of Paradise, &c. Note, that this work may be done also with ☿ dissolved in A. F. and precipitated with spirit of salt; the Calx dryed and united with Calx of ♃ and ♂, and thence the volatile crystals, or butter extracted, wherewith you may proceed as was said: Or, you may make also a ▽ of Paradise, made with ♓ *per deliquium*.

The Antimonial ▽ of Paradise,

or the Hermetick Coelestial Eagle with two Heads.

Extract a butter from equal parts of ♂ and ☿ sublimate: Dissolve this butter in the air in ♈, ♉, and ♊; put the liquor into a glass cucurbite with its head and Recipient, lute well all the junctures; excite the *Archaeus* which is in him, by a very gentle heat in ashes, by a Lamp △, which will maturate the matter in the space of fifteen or twenty days: Then drive up its rays into the head, which will be seen corporal in the Recipient in the form of a clear ▽. This ▽ is all fiery, and is the Coelestial Eagle with two heads. Put it into a cucurbite, and deflegm it with the same Lamp △, and there will remain in the bottom of the cucurbite the mineral gluten, or the viscous ▽, which does not wet ones hands. You may prepare medicine of this Coelestial AEthereal ▽ thus: Put two drops of it into ℥iv. of S.V. itwillturnas white as milk. This Medicine Cures the Dropsie, the Epilepsie, Madness & etc. The dose is from ℥ij. to ℥ ß. Now, if you would

214

have the powder of projection, you must digest the gluten *per se*, as was said; or (which is better) add a tenth part of ☉ in leaf, and digest or draw the red and white Oil, and proceed as in the former process, and you shall have a medicine both for man, and for metals.

Water of Paradise of Venus and Mars,

or ♀ and ♂ Captivated, whence comes Cupid, or the Solar Panacea.

Although these metals cannot take the Mercurial ▽ nor give volatile crystals, as ☽, ♃, and ♄ do, because they are very Mercurial, and the former almost all Sulphurous, nevertheless you may do it thus: Dissolve ♀ and ♂ (each by itself) in the salt *Androgine*, which has but little Sulphur, to the end, that it may dissolve more easily: Then make a Lixivium, which precipitate with your liquor of ♄; dry the precipitated Calx, and sprinkle it with a good deal of spirit of salt; then mix it with ☿, and distill volatile crystals thereof; with which proceed as before. The ▽ of Paradise is made of ♀ only, and is called ▽, or ♀ his *astrum*.

It cures the Pox, Gout, &c. The gluten of these metals is digested either *per se*, or with a solar

215

ferment, as before, into a *Panacea*, which is a wonderful medicine, and will cure maladies in men and metals.

The Thrice Noble Water of Paradise,

Or Apollo Medens.

Distill the fiery and volatile crystals from ☿ of ☽, with Calx of Jupiter, which keep. Dissolve ☉ in spirit of salt, which join with equal parts of your crystals; digest, and then distill, the spirit of salt will come over immediately first, then will follow the red crystals: Expose this terrestrial ☉ to the Coelestial, that it may satisfie itself with its solary rays, and then dissolve itself into a liquor, which will be a magnet and an Amaranth AEthereal and immortal. Put this solary and lunary liquor into a glass cucurbite, and distill with a Lamp △ this noble, metalline, radical moisture, those invisible rays of the ☉, or this ▽ of Paradise, during forty or fifty days. This ▽ is ▽ of Nature, an excellent attractive, and its power is ineffible. This ▽ drives out all Maladies, and comforts Nature, and is a royal medicine; for 'tis the astrum of the ☉, or a ☉ between the Terrestrial and Coelestial ☉. Of this is *Apollo*

216

furens; for its rays, or its ▽ kills ☿, which they convert into true ☉, as also all other metals. In this liquor you may dissolve ☉ if you will, but it will not be necessary; for when it is freed of its flegm, the solary gluten remains, which you may digest *per se*, until it acquires a purple colour. Thus ☉ is exalted to make a tincture. The ▽ of Paradise is the AEthereal Aurum potabile; dissolve two drops thereof in ℥iv. of S.V. the dose is ℥ij. This is the ▽ of Nature, which is multiplied *ad infinitum* by new addition of the gluten, &c. Note, that when this *Panacea* is fixed, it is the *Panacea* of *Panacea's*, which cures maladies, both in men and metals.

Note, that this ▽ of Paradise converts all metals into ☉, if you digest their plates in the same; yea, one drop thereof penetrates a plate of ☽, and transmutes it into most fine ☉. There is also made another *Apollo Medens*, which is joined with *Spiritous Regulus* of ♂, to wit, the flowers reduced, or fiery Regulus, and conjoyned in the Sulphurous *Sal Enixe* and both precipitated into an aureal antimonial *Panacea*. But this *Panacea* is not comparable to the other. *Apollo furens* is the same ▽ of Paradise, the which are the invisible solary rays, by which the volatility of ☿ is killed, and is converted into ☉; and the same it is with the solary ▽. *Apollo Moriens* is the Eclipse of the ☉ in the above-mentioned fiery and Aethereal Menatruums: For in all sorts of Menstruuma it putrifies, grows black, and maturates in the space of fifteen days: But after that, it resuscitates before the Judge *Apollo resuscitans.*

An Unheard of Arcanum,

or new and Unheard of Lunary, wherewith is made the Elixir, or Metalline Stone.

Dissolve what quantity of ☽ you please in the *Sal Androgine* in four hours' time half your ☽ will be dissolved into a very red salt, pour it into a vessel of copper, then make a Lixivium, which filter, reduce into a body what remains with ♄, and re-dissolve it in new *Sal Androgine* as before. Reiterate this, till all your ☽ goes through the filter with the Lixivium, and you will be sure to have a ☽ altogether spirituous and volatile, which you will find to be true to your loss, if you precipitate it with an acid liquor, and reduce it with ♄; for it will all fly away at the Coppel; the same will happen if it be attracted by plates of ♀.

These two effects have happened to me by
inadvertence.

Note, that this spirituous ☽ is a potential
and spirituous ☉, as you will find, if you rejoyn
it with its body at the Coppel: There is nothing to
say to that. Note, that the corporal ☽ which is
added, retains all what is of the Nature of ☉,
which renders it afterward in the separating ▽[16].
Therefore, take all these filtered solutions (which
are yellow if the Lavers be made with Odor of
Metals[17]) and precipitate them totally into a lunary
sulphur of a golden colour, adding a sufficient
quantity of that which precipitates it: That which
precipitates it is of our invention, and is of the
Saturnian juice, which swims upon the ☿ of ♄, when
its solution is repercuted by the salt ▽. Dry this
lunary golden precipitate gently, and mix it with
ana of Calx of ♃, made *per se* in the ▽; or if you
will, you may draw the butter or fiery crystals with
ana of ♂; the crystals are resolved per se in the
air. And with this unheard-of magnet are
miraculously attracted the influences of the stars,
or the ▽ of Nature. This is chiefly done in the

---

[16] This symbol was used by Hans and in the printed 1682
edition. I believe it should be △. -pnw
[17] This parenthetical comment makes no sense to me. It appears
here as it was in the 1682 edition. -pnw

belly of ♈; that is to say, in the months of April and May. Note here a very great secret, which is, that there is no flegm attracted by this magnet, but only the pure nutriment of life, or the fiery vital Viand, which is hidden in the centre of the air; which you will find true, if you put some waterish part into this liquor; for you shall see that it will not mix with it in any wise, but will swim upon it in an heterogene form, as Milk: You must further separate this liquor, which is the simple lunary, in which ☉ is easily dissolved: For from this corporal lunary you must have a spiritual and unheard-of lunary. Put then this liquor in a glass cucurbite with its head and Recipient, and digest in ashes with a very gentle heat by a Lamp the space of one Philosophical month. Nothing will distill over during the first fifteen days, or more; but it will become a red Sea, and the matter will maturate, and after that, you shall see that by this gentle heat the metalline soul will mount invisibly upon the wings of the wind, or the spirit of the world, and will fall into the Recipient in the form of tears, which are the tears of Diana. This liquor is much more precious than pure ☉, and of very great virtue. Continue the dissolution, whilst the *Archaeus* of Nature chases it, which is done in fifteen days at the farthest. In this operation is done, what Hermes said, thou shalt separate, the subtle from the spirit gently, and with great dexterity. This distillation is altogether Natural, and is perfected by the only Archaeus of Nature. This liquor is the Spiritual Lunary which contains in itself Body, Spirit, and Soul; 'tis the ∇ of Paradise, the Lunar Sphere, the Metalline Fountain, and the Universal Metalline Menstruum. It is a most certain anti-epileptick and Cephalick: If one or two drops of it are mixt with $\mathfrak{z}$iv. of S. V. all will become like Milk. For it is all $\triangle$; which changes the moist Element of the S. V. as being contrary to it, or at least not connatural.

To make the Metalline Stone

per se of this Spiritual Lunary.

Take this liquor, and put it into a small glass cucurbite; leave it uncovered, evaporate it in ashes with a gentle Lamp $\triangle$, to the end, that if there be any moisture from the air, it may exhale, and there will remain in the bottom of the cucurbite the metalline gum, the lunary gluten, the Azot, &c. which will liquefy at the least heat, as butter, and will congeal by cold. Put this gum into a Matrass, which seal hermetically, and digest per se, it will become black, and after white, and then it is the white stone; then by increasing the △, it will become of a citrine colour, and red, without a solary ferment: And the King is made of the Queen, or the immersion of ☽ into a solary tincture. But for to shorten the work, add unto this gluten a tenth part of ☉ in leaf, or Sulphur of ☉ made spiritual by the Sulphurous *Sal Enixe*, and digest as was said: The augmentation of this stone is by addition of new metalline gluten. Note, that this spiritual lunary tinges ☿ into true ☽, if you digest it therein; also a plate of ♀ is perforated by putting one drop of this ▽ upon it. Note also, that when you have distilled the lunary, that which remains is an eternal magnet. To that effect, resolve it again in the air, and manage it by a Lamp as before. Then distill an AEthereal Liquor, which is yet impregnated with a lunary soul, and distills into the Recipient, and then goes anew into a gluten: And this is done *ad infinitum*, Note also, that that which remains, may be distilled, and you shall have first, a white lunary Oil (which is the philosophers Oil of Talc; for the true Oil of Talc is the lunary coagulated *per se* and fixed into a white stone, which is fixt and soft.) Secondly, you shall have the red Oil by augmenting the △. If you will make the stone of these matters, take of the red Oil one part, and of the white Oil four parts, and eight parts of the lunary reduced into gluten: Put this into a Matrass, and digest until all be fixed into white, and after by continuing become red. This medicine ought not to be fermented; for it is the true metalline soul, reduced into a tincture. This last digestion must be in an Athanor with a charcoal △.

(Hartman) These waters of Paradise and Glutens, and c. were given to Sir. K. (about eight or nine months before he died) by a French Gentleman, a great Scholar.

Monsieur Barkly's fixation of Common Sulphur,

and the Tincture thereof, which is an Excellent Medicine in all affects of the Breast and Lungs.

Take flowers of Sulphur, or Sulphur pulverized very subtle; put it into a Matrass, and pour upon it so much spirit of Sulphur *per Campanam*, as may cover it the breadth of three fingers. Lute the Matrass well, and put it in digestion for fifteen days, or three weeks, or so long until the flowers of Sulphur come to be very black: Then distill off all the spirit of Sulphur to dryness; break the Matrass, and take out the Sulphur, which pulverize again, and put it into another Matrass, and pour upon it the spirit

223

of Sulphur you distilled off, and distill as before to dryness. Repeat this twice more, which makes three cohobations in all without the first distillation. Then take your black and fixt Sulphur, and reduce it to a very subtle powder, and put it to reverberate in a glass oven the space of a fortnight or three weeks, it will change its blackness into white, and after yellow, and at last come to be of a reddish brown colour. The tincture of this red fixt Sulphur, is extracted with spirit of salt well rectified.

He made thus his spirit of salt for this: Take salt lb j. dissolve it in five quarts of fair ∇, and filter it; put it into a cucurbite, and pour upon it by little and little lb j. of good Oil of Vitriol, and join the head and Recipient; when it is all in, it will begin presently to distill over cold: Set it in sand, and with moderate heat drive over as much as will rise, which rectify from the flegm: There will remain in the bottom of the cucurbite a wonderful salt, that is exceeding fusible.

After he had extracted the tincture, he distilled away all the spirit of salt, till the tincture was dry: Of this he gave three grains for a dose, and found it a great diaphoretick, but it was somewhat rough and sharp in the stomach: Whereupon he dulcified it by several ablutions in fair ∇;

then gave the same dose, and it wrought excellently well in all colds of the breast and lungs.

(Hartman) This relation is of Sir K. Digby.

The Countess Of Kent's Powder,

as it was prepared by Sir Kenelm Digby's Order in his Operatory.

Take $\mathfrak{Z}$iv. of the black ends of the shares of crabs, the sun being in the sign of Cancer, crabs-eyes, fine pearls and corals prepared, of each $\mathfrak{Z}$j. yellow amber $\mathfrak{Z}$ ß, Roots of *Contrayerva*, *Virginian* snake-root, *ana* $\mathfrak{Z}$vj. Oriental *Bezoar* $\mathfrak{Z}$iij. of the bones that are found in the hearts of stags Θiv. Reduce all into a subtle powder; moisten the crabs claws and crabs-eyes, and the powders of pearls and corals with a little juice of lemons, to make them ferment a little: Then the next day mix all well together, adding $\mathfrak{Z}$j. of tincture of saffron, and pour upon the mass (when you incorporate it) three or four spoonfuls of spirit of Honey, or instead thereof you may take jelly of Hartshorn, and jelly of the skins of vipers dryed in the shadow. Then add to this composition $\mathfrak{Z}$j. of *Trochisque* of vipers; grind it all well together to make it well incorporate: Then make it up into balls, and let them dry, and keep them for use.

This Powder is a most excellent remedy in all epidemical distempers, all malignant, spotted, and purple fevers; to drive out the small-pox and measles. It is sudorific, and resists all corruption, and is admirable in a surfeit. It drives the venom from the heart, and hinders the vapours to fly up into the head and brain. It drives out by transpiration all bad humours, corroborates and strengthens nature. The dose is from six to twenty, or twenty five grains. In an extremity of the Plague, one may take from thirty to forty grains.

(Hartman) Sir K. D. had the Powder always ready by him in his closet; and I remember that many persons of quality sent to him for some of it when any of their children had the small-pox or measles; and never any did miscarry of all those that took it. It is also excellent against the biting of mad dogs, stinging of vipers, and other venemous beasts.

A very Efficacious Remedy against Epilepse,

or Falling-Sickness, wherewith Sir Kenelm Digby Cured a Ministers Son, named Mr. Lichtenstein, at Francfort in Germany, in the year 1659. to which I was an Eye-Witness.

Take of the skull of a man that died of a violent death, of the parings of nails of man, *ana* ℥ij. Reduce this to a subtle powder, and grind it upon a marble stone; then take Polypody of the Oak very dry, ℥ij. Misletoe of the Oak, gathered in the wain of the Moon, ℥β. Misletoe of the Hasle-tree, Misletoe of the Tile-tree, of each ℥ij. Piony-root ℥β. Reduce all into a subtle powder: Then take ℥vj. of sugar, boil it to the consistence of rose-sugar; then mix all the powders with it, and stir them well together over the fire that they may well incorporate together: Then take it from the fire, and make it up into little tablets of about a dram apiece; whereof give one in the morning fasting, and two or three hours after dinner, and another two hours after supper: Continue this whilst the tablets last.

Another for the same.

Sir Kenelm Digby relates, that in the year 1663, the Lady Warwick told him, that a daughter of her husband's elder brother had the falling-sickness in the greatest extremity, so that she fell like a log seven or eight times a day without any motion. They had put her into the hands of the ablest physicians in England, who in effect could do her no good. A gentlemen, one of their neighbours, undertook to cure her, and performed the cure thus: Take true misletoe of the Oak, the leaves, the berries, and all the tender branches; dry it gently in an oven after the bread is drawn; then reduce it to a fine powder, of which give as much as will lie upon a shilling for one of ripe years; for middle aged, a six-pence, for a child, a groat: Give it mornings and evenings in cowslip-water three days before, and three days after the full of the Moon. Repeat this remedy for some months together. This cured also my Lord Herbert's son, and many other persons of quality. The best time to gather the misletoe of the Oak, is in the month of September, when it bears berries, and in the waning of the Moon.

Preparation of the Silver Pills against Dropsie,

as they were prepared by Sir
Kenelm Digby's Order in his Operatory.

Take refined ☽ ℥j. dissolve it in ℥iij. of the best spirit of Nitre in a Matrass, then evaporate away all the spirit of Nitre to dryness in a low cucurbite, or in some other fit vessel; then dissolve the matter in a sufficient quantity of rose-water, filter the dissolution through gray paper, and evaporate it again to the consistence of a dry salt as before. Then take ℥ij. of fine salt-petre, dissolve it in rose-water, filter the dissolution, and evaporate it in a large wide vessel of glass, to the consistence of a salt. Then mix the ☽ and this salt together, and put them in a large glass, pouring upon them so much rose-water as will dissolve them into a greenish liquor: Then evaporate it in sand to the consistence of a white salt; then take it out of the sand, and being quite cold, put it into a glass or marble mortar, and put to it ℥ij. of fine wheat-flower; grind them well together, then add so much rose-water as will make it a mass fit for pills: Then make it up into pills of the bigness of peas, put them between two papers, and let them dry in the shadow, and they will be of a purple colour; keep them in a wooden box.

Directions for the Use of these Silver Pills.

They are a specifick against the dropsie, the patient is to take one of them at six or seven of the clock in the morning, taking some broth about two hours after it with eight or ten drops of spirit of salt in it. Their operation is by stools, and by urine; you must continue it until the cure be perfected. Note, that if the patient be weak, he must take the pill but once in two days, and in all broths and drink, he ought to take some dose of spirit of salt, as is said above. If there be need of sweating, you must use some dry stoves, and give him always of the following salts: Take salt of urine, salt of worm-wood, *ana* ℥ij. add half a scruple of Oil of Amber, and as much of spirit of urine, with ℥ij. of fine sugar; mix all well together in a glass or stone mortar, whereof give Ɔiv. for a dose in half a glass of white-wine when the patient is sweating in the dry stove, and not in a bath of water: And every third day you must repeat this remedy, and he will be cured within three days. The evacuation is by abundance of sweat and urine.

(Hartman) I cannot omit to relate here a story, which I have often heard Sir Kenelm Digby tell concerning a famous cure of a desperate dropsie, done by Dr. Farrar upon an eminent Lord, who was over-grown with the dropsie, his belly and stomach swelled to a prodigious bigness, and was given over by the ablest physicians as incurable. Sir K. D. made the bargain between the Lord and the Doctor, who was to have five hundred pounds for the cure:

But when the Lord was cured, he would give the Doctor no more than three hundred pounds, saying, that five hundred pounds was too much money, and that all the ingredients he used could not stand him in twenty shillings. The remedies were thus: Having first well purged the patient with some fit purge (as of Jallap, Manna, Sena) to carry away wary humours, he gave him the following broth. A moderate broth was made of mutton, chickens, and capon, or hen, but not veal; the broth was not strong of the meat, nor too weak, but such as the patient might drink all the day, for he was to drink no other liquor; they made but about a pottle of broth at a time, for it would not keep: And for this quantity they took a gallon of water, into which the doctor put above a handful of garlic, and rosemary, pennyroyal, thyme, sweet marjoram, fennel-roots, parsley-roots, as also currans, and a sufficient quantity of salt. And after some days taking the broth, they put into every draught of the broth (the patient took) above a spoonful of the crude juice of garlick, stamped and pressed out. But if you cannot bear always to drink this broth, then use the following decoction: Take Sarsaparilla $\tilde{3}$xij. China-roots $\tilde{3}$v. Sassafras $\tilde{3}$iij. Cut all these very small, and pour upon them spring-water, to three fingers breadth above the ingredients, and let them infuse over a soft fire the space of four hours; then throw away this water, and stamp the ingredients in a stone mortar with a wooden pestle: Then pour upon them ten quarts of fountain-water, and boil it in a vessel close stopped, till four quarts of it be consumed: Of this decoction let the patient drink, without any other drink but the garlic broth.

Another Drink.

Take all the aforesaid ingredients, in the same manner prepared and stamped: Then take a clean vessel, and fill it with beer, then put the ingredients in a bag, and hang it in the beer; $\tilde{3}$j. of the ingredients is sufficient for a quart of beer. Either of these drinks is only is case the patient cannot bear the use of the garlick broth, which alone will dispatch the cure much the sooner; and this course of the garlic broth is for all obstructions, and superfluity of cold, raw humours, clogging the brain, or any other part, as well as for the dropsie. To strengthen and secure the liver, use the following electuary. Take of powder of Turmerick a sufficient quantity, make an electuary of it with sugar, and to every ounce of it add three drops of Oil of Anniseed, made by distillation; and if you put a little *Ambergrease* to it, it will be the more strengthening. Take of this electuary two or three times a day the quantity of a hasel-nut; take not above ℥j. in a day.

Besides this, to strengthen the stomach, use the following stomacher: Take worm-wood, marjoram, rosemary, rue, *ana* one handful; cloves, cinnamon, mace, *ana* ℥j. bruise these spices, and mix them with the herbs; of these make a stomacher, and apply it: And you may likewise anoint your stomach, and region of the liver with Oil of nutmegs and Oil of roses.

I heard Sir K. D. say, that after twelve or thirteen days, the patient begun to piss in great abundance, and so stinking, noisom, roping matter, that the nurse which emptied the pots, was hardly able to endure the stink and noisomness of it. And he continued the diet till he was perfectly cured.

Another Experimented Remedy for the Dropsie,

whereby several Persons have been Cured, as I have been assured.

Take the root of heath, scrape off the first bark, which throw away, then peel off the next rind, and fill a glass or a bottle with it loosely, then fill it up with white-wine, and let it stand to infuse overnight, and the next morning drink half a pint of wine; and so continue until you are cured.

Another Excellent Remedy against the Dropsie.

Take spiritual Oil of Salt, mix with it so much flowers of Sulphur, that it become like pap, which distill in a retort in sand, and you shall have a liquor as white as Milk, which is excellent against the dropsie.

The Copy of a Letter from Abbot Boucaud

from Paris to Sir K. D. wherein he relates in what manner he Cured himself of the Stone, and of a Quartan Ague.

Sir,

I do not tell you that I have been sick, (and that I am so still) to excuse myself for having so long deferred an answer unto your last two letters, &c. It is true nevertheless, for I have laboured under diverse distempers; but among the rest, I have been ill of the stone, and have had a Quartan Ague: I believe you will not be sorry to hear how I cured myself of both without the help of any physician. For the stone I took twelve grains of the salt made of the stones which were taken out of men; I dissolved the said salt in a little water, and then I put all into a glass of white-wine, and drank it off, and walked about my chamber near two hours, at the end whereof I had a great need to make water, and I voided (with violence) a large glass full of gravel, which was so gross, and so rugged, that it caused me to void near a pint of blood; the same thing happened to me three times, and every time I voided blood, which made me judge that I should have taken less of the salt; yet I took it but once, but I felt a great pain and heaviness in my reins and kidneys. The said stones were calcined in a potter's oven, and after they were calcined, I extracted the salt out of them with distilled rainwater: The feces I calcined again, and extracted the salt as before, which I repeated so often, till the said stones yielded no more salt. Note, that to make this salt for a man, you must take the stones taken out of men, and for a woman, those that are taken out of women. And thus, was the first cure performed.

As for the Quartan Ague, without having been purged, or let blood, at the fourth fit I took a glass-full of the water of green wall-nuts, which I had distilled in their last season: I took it as soon as I perceived the least symptom of the fits approaching; I went to bed, and caused myself to be well covered, and slept, and had no fit at all that time, nor ever after.

The water I distilled thus: I took green wall-nuts and beat them in a stone mortar, then in a cucurbite in B. M. I distilled the water from them, which I cohobated twice upon fresh wall-nuts. Then having calcined the three Marcs or Caput Mortuum, I extracted the salt out of the ashes; this salt I put into the distilled water. I thus, Sir, I have given you account how I went to work.

A Process: how to make a most Excellent

Oil of Sulphur in Abundance;

sent also by the said Abbot Boucaud to Sir K.

Take an Earthen pan of stone-ware, in the midst thereof lay a piece of brick, upon which set an Earthen Poringer full of Sulphur grosly beaten; then put fair water into your pan, but no so much as to touch the said Poringer: Then kindle the Sulphur, and cover it with a bell, so that the bell touches the water, and that the fumes may not come out, but may condense and run down into the water, which afterwards must be separated in B. with a moderate heat. To set the Sulphur on fire, you may put into it a square or round piece of iron made red-hot in the fire.

(Hartman) In my opinion, if the bell touches the water, and it has no hole at the top, so that the Sulphur will have no air, it will not burn; I judge the best way to be thus: Let the Poringer stand in the water, but not so deep, as that the water bear it up, and make it float; if it stands half way in the water, it will do, for the weight of the Sulphur will keep it down, and the heat of the Poringer will heat the water, and the vapours and steams thereof will mix with the fumes of the Sulphur, and make them condense the better, and so distill down together into the water. The bell should be such a one as is now in use, with a long neck, and a hole at the top, which should not touch the water nor the pan, but it should be suspended in such manner, that there be some distance between the brim of the bell and the sides of the pan.

A subtle Volatile Water from Sulphur,

which will Dissolve ☉.

I am told by one who has done it, that when you go to sublime flowers of Sulphur, if you give very gentle and moderate fire, and be very attentive, there will come over first, before any flowers sublime, a little very volatile, but altogether insipid water, which he said, will dissolve ☉: It is much more volatile than any S. V. A glass full of it will presently vanish away, if you hold the glass unstopped upon your hand, by the warmth of it.

(Hartman) This relation is of Sir K.

If you would save this water, you must have a glass head upon your last subliming-pot, or a ludel, wherein you sublime your flowers of Sulphur, and instead of a vessel without a bottom, as that for the flowers of Antimony, you must have one with a bottom, and without a hole on the side to put in your Sulphur, and then two Aludels besides the said vessel, and the glass-head will be sufficient for subliming the flowers of Sulphur.

By means of the glass-head you save also the vinegar of ☿ in subliming the flowers, which I have done several times; but I used not above three Aludels one upon another, besides the glass-head.

An Excellent Essence of Sulphur

for the Breast, and for the Lungs.

Take Sulphur one part, brown sugar-candy two parts; pulverize them, and mix them well together, then put it into a retort of such a bigness, that two third parts thereof may remain empty. Then distill in sand, giving very gentle fire at first; you will have a whitish liquor, which keep for use.

(Hartman) This was given me by a physician at Paris, who told me, that a Catarrh failing upon his Lungs, which obstructed his Lungs, causing in him a great Fever, he cured himself with this essence, taking this, thirty or forty drops of it in some broth. He told me also, that it was of great effect in Asthma, Phthisick, old and inverterated Coughs, &c.

An Excellent Elixir of Sulphur.

Take juice of Licorise, Confection of Alkermes, Roots of Elecampane, *ana* ℥vj. *Alipta Moscata* ℥iv. Myrrh, Saffron, *ana* ℥j ß. Mastick, Benjamin, Cardamoms the less, Cinnamon, *ana* ℥j. Sugar-candy ℥ ij. Powder what is to be powdered, then mix them together, and add rectified S.V. so much as to make it into a paste; then put it into a circulatory vessel, and pour upon it so much spirit of Sulphur, as may cover it the breadth of four fingers: Digest it forty days, then decant the tincture, and pour upon the remaining matter fresh S. V. to extract another tincture. Then mix these two tinctures together, and keep them for use.

This tincture is a very great pectoral, and a precious remedy in all affects of the breast and lungs. It is excellent against catarrhs, old and inveterate coughs, the phthisick, asthmas; it cherishes and comforts the heart, and is good against fainting and swooning fits, preserves from putrefaction; it is Anodyne, Cephalic, Analeprick, Alexipharmack; and as the author said, preserves health, prolongs life, and keeps back gray hairs, by strengthening natural heat. It is to be taken in some pectoral water or syrup; the dose is so much as renders the vehicle of a grateful acidity.

Lac Sulphuris.

Take of Sulphur in powder one part, and of Quick-lime two parts, mix them, and put them into an iron pot, and pour thereon a good quantity of fair water, let it boil until three parts of the water are consumed, and that the liquor be as red as blood by the dissolution of the Sulphur; then strain it whilst it is hot, and let the strained liquor stand to cool: Then precipitate with vinegar, then let it settle, and having poured off the clear, edulcorate the residue ten or twelve times with warm water, the last time with rose-water; then dry it gently, and keep it for use.

It is a true remedy in all affects of the breast and lungs; it is given with great success to those that are troubled with Catarrhs, Rheum in the head, Asthma, Phthisick, Coughs, &c. It promotes expectoration; it hinders the defluxion to the joints, it prevents and disperses the windiness of the stomach and bowels, and cures the cholick. The dose is so much as may change the vehicle white; the best and fittest vehicle is the spirit of *Lignum Cassiae*, or Cinnamon; taking it twice a day, in the morning fasting, and at night.

You may make a very good spirit of *Lignum Cassiae* thus, which is a much finer spirit than that of Cinnamon, and much better for this use. Take *Lignum Casiae* ℥iv. bruise it well, then pour upon it three quarts of *Malaga* Sack, stop the vessel close, and let it stand to digest for three or four days, then distill it in a Limbeck, or in a glass cucurbite, distilling it off all together, as long as it comes with vigour, and you shall have about three pints and a half of very good spirit: Thus I make it. But if you will have it richer of the wood, put this liquor upon fresh *Cassia*, and digest and distill as before. Repeat this till it is as strong as you desire. You may if you please separate the runnings so as to have some of such strength as you wish.

A Great Diaphoretick of Antimony.

Take good Antimony Mineral in subtle powder lbj. mix it with lb ℔ of ☿ sublimate; put this mixture presently into a retort, leave the retort for some time unstopped before you distill it, for then you shall have more butter than if you distill it presently. Then distill a butter from it according to Art, giving strong fire at last, so that the bottom of the retort may be red-hot; part of it will come over in butter, and part will sublime in Cinaber, very hard; if you leave this butter for some time exposed to the air before you rectify it, you shall have more liquor than if you distill it presently; rectify this butter, then melt it again, and pour it into a clean retort, and pour upon it by little and little good spirit of Nitre, continue pouring on the spirit of Nitre until the ebullition ceases: Then distill it with a gentle fire in sand, giving strong fire at last, so that the bottom of the retort may be red-hot; then let it cool, break the retort, and take out your matter, which will be very spongy, and of a yellowish colour; pulverize and edulcorate it several times with warm water, then dry it gently; reverberate it for an hour between two crucibles well luted together: Then grind it again to a subtle powder, which put into an Earthen Poringer, and pour upon it rectified S. V. that will burn all away; fire it, and whilst it burns, stir it continually with a silver spoon; the S. V. being burned away, the powder will remain dry; grind this powder again, and mix it with ℥viij. of Antimony Diaphoretic that has been calcined three times with Nitre, grind them well together, and put them into a retort, and pour upon them ℥iij. ℈. of good spirit of Nitre; put the retort in sand, and let it stand thus four and twenty hours; then distill with a gentle fire to dryness: Break the retort, and take out the matter, which grind and edulcorate with Carduns Water warmed, then spread it upon gray paper, and let it dry of itself; Then grind it to an impalpable powder, which put into a Poringer, and pour upon it S. V. so much as may cover it a fingers breadth; let it stand thus for five or six hours, then fire the S. V. upon it, and stir it continually with a silver spoon whilst it burns, then grind it again, and put it into a vial, stop it close and keep it for use.

The manner of using this medicine is thus: Take fifteen grains of it for three mornings together, mixing with it some conserves or sweatmeats, and take it upon the point of a knife, then drink a glass-full of the sudorific decoction after it warm. Then take twenty grains for three mornings more; then fifteen grains again for three mornings more. It is an excellent remedy to cure the gout, dropsie, palsie, the venereal disease, the evil, leprosie; it purifies the whole mass of blood, and is good in all scorbutick distempers. Note, that before you use this medicine, you must prepare the body before with some fit purge, according to the constitution of the patient. Those who are careful to preserve their health, and to keep it in good state, may take this powder in the spring, at the falling of the leaf, having first purged once or twice; then take the powder with the sudorifick decoction for nine days together, as was said, mixing the powder with a dram of confection of Alkermes. It powerfully resists all corruption, dries up all superfluous moisture in the body, and is a true concretive of blood.

The Sudorifick Decoction.

Take Lignum Guaiacum ℥iv. Sarsaparilla., Sassafras, *ana* ℥j. infuse them in three quarts of fountain-water for twenty four hours; then let it boil gently for three hours.

A Most Excellent Medicine

Against All Sorts of Agues and Fevers, &c.

Take of the starr'd martial Regulusof Antimony lbj. Mercury precipitate lbj. ß. pulverize and mix them well together, then put them in a retort, and distill in sand as you do butter of ♂; then rectify this Oil or butter once or twice, casting away the feces: Then put it into a new retort, and pour upon it spirit of Metheglin; distill and cohobate four or five times to make the Oil sweet, then pour S. V. upon it, and abstract it to the consistence of an Oil. This is a precious remedy for the cure of many diseases:

It is of great power and efficacy to cure all sorts of Agnes, Quotidians, Tertians, and chiefly Quartans. It operates by a gentle vomit in some persons, and in others it gently purges without vomiting, and in some it gently operates both ways:

It has virtue to eradicate totally both root and seed of the distemper. The dose is from six to twelve drops, in some fit vehicle. Note, That having separated the spirit of Metheglin, if you acuate it with spirit of Vitriol, it is a great Diaphoretick, far beyond all others. Dose is from half to one whole spoonful in some fit vehicle.

A Precious Oil of Antimony.

Take Antimony calcined, as for making the glass of ☿, lbij. ℥xij. Sugar lbj. Mix them well together, and put them in a retort: Distill in sand, first, will come a flegm, and afterwards a pure dark-red Oil, which keep for use.

This is an admirable remedy against the stone and gravel, the dropsie, epilepsie, asthma, quartan agues, and all sorts of fevers, the plague, and all malignant fevers, and epidemical distempers, and leprosie; and being outwardly applied, it cures, heals, and dries up all inverterate wounds and ulcers. The dose is four drops in wine twice a day.

A most Excellent Panacea of the true Sulphur of Antimony.

Take Lees of wine, which you may have of the wine-coopers when they have pressed them out, break them into small pieces, let them dry, then burn them to ashes: Take of these ashes, of Quick-lime, and Nitre, *ana*; made a Lixivium thereof with warm water, then filter it: Then take Cinaber of Antimony, which is found in the neck of the retort when one made the butter of Antimony; pulverize it, and boil it in the afore-said Lixivium for the space of four hours; pour off the Lixivium from the Quick-silver into another vessel, which lean on the side, that the red Sulphur may settle; then edulcorate it with hot water, and dry it gently; so have you the true Sulphur of Antimony. Take of this Sulphur, and of Regulus of Antimony, *ana* $\mathrecipe{z}$j. Oil of Sulphur per Campanam, or rectified Oil of Vitriol $\mathrecipe{z}$iij. Mix all well together, and put it into a small retort, digest it in horse-dung, or if you will, in some other gentle heat for eight or ten days. Then distill it, and cohobate the distilled liquor upon the mark three or four times; then increase the fire to the highest degree, which continue for twelve hours, to force all over, and the matter will be fixt; then break the retort, and take out the matter, which pulverize, and edulcorate it with rose-water; then dry it gently upon a gray paper, then reverberate it for four or five hours. Then take ℥j. of this powder, and of salt of red Coral ℥ij. grind them well together to a very subtle powder.

This is a universal medicine to purify the whole mass of blood, and to root out such distempers as proceed from the corruption thereof, and are curable by sweat. It cures all stubborn, malign, and chronick diseases: It cures the venereal disease, the most inveterate; the leprosie, the evil, the scurvy, the plague, and all epidemical diseases. The dose is from ten to thirty grains.

The order of using this medicine is thus: First, purge the patient once or twice with fit purges, then rest three days, then purge again; then begin with ten grains of the powder, which continue for three times, mix the powder with some fit conserve, and give it upon the point of a knife, in the morning in his bed, drinking a glass-full of a sudorifick decoction after it, made hot; let him keep his bed for an hour or two, then let him be rubbed with warm clothes, and the sweat being quite over, let him rise, and eat of good wholesome food, forbearing to eat of salt meats, salt-fish, sallet, milk, butter, or cheese, or raw fruit. Then for three mornings more give him twenty grains, and then thirty grains for three mornings more; then come again to twenty grains for three mornings more.

A great Febrifuge.

Take mineral Antimony very clean, that has never been melted, $\text{\textasciitilde}vj$. and as much salt-petre, pulverize them finely, and mix them well together; then put them into a strong crucible, which cover with another crucible that has a little hole in the bottom as big as a pea: Then put this crucible into your furnace, and let the fire kindle of itself; which increase by degrees, the matter will fulminate; when you see that no more smoke comes out of the little hole of the crucible, take it out of the fire, and take out the matter that remained in the crucible, which pulverize very finely. Then take three ducats of Gold, and six times as much in weight of the afore-said powder; melt the powder first in a crucible, then put into it, one of the ducats, stirring it until it be melted, then put in another ducat; and so continue until you have put in all your ducats one after another: When all is melted and well incorporated, let it stand in good fusion for half an hour, then take it out, and let it cool: Then break the crucible, and take out the matter, which pulverize subtily, and mix it with equal weight of ☿ sublimate, also in fine powder; put them into a retort well luted, put it into a furnace, and fit a Recipient to it full of water, so that the nose of the retort may enter into the water; leave the junctures unluted:

Give a gentle fire at first, which augment by degrees; part of the matter will distill into the water, but the greatest part thereof will stick to the neck of the retort, which you may draw out with an iron hook into a bason full of water: When you see that nothing more comes over by the last degree of fire, let it cool; then break the retort, and take out all the matter that is sublimed about the neck of the retort, and put it into the water in the Recipient, as also that in the bason; let the water stand to settle, then decant it, and keep it. It is excellent to cure all sorts of old and inveterate ulcers, &c. Pour fresh hot water upon the residue, and having shaken it well together, let it settle; then decant, and put on more water. Repeat these solutions seven or eight times; then separate the from it with a Quill, and put the powder into fresh hot-water; let it stand thus until the next day, then repeat the edulcorations as before, which con-tinue for six days, then edulcorate the last time with cold water; then dry the said powder, and keep it for use. The dose is one or two grains for children; and for persons of riper years, from four to six or seven, according to their strength and constitution, putting the powder over-night to infuse in two or three ounces of white-wine; the next morning strain the wine, and let the patient drink it, and half an hour after he may drink it, and half an hour after he may drink some warm broth or Poffet: It may also be given in substance. It operates by a gentle vomit, and by stools. It has been experimented, and found very successful and effectual in the cure of all intermittent fevers, and in the gout, as also in the venereal disease. Out of the Caput Mortuum you may reduce the greatest part of the gold.

This is a Mercurius Vitae of a singular preparation; it is not white like the common, but of a brownish gray colour. It appears by this to have some of the ☿ in it, that when you rub gold or copper with the powder, it will make it white, which common *Mercurius Vitae* will not do.

Another great Febrifuge,

which is said to be Riverius his Febrifuge.

Take ☉, dissolve it in A. R. and glass of Antimony, dissolved in A.F. *ana* ℥ß. washed and purified ℥iij. Dissolve it in A. F. Then mix the three dissolutions together, and put them into a

251

cucurbite, and distill in sand, and cohobate the distilled liquor eleven times upon the remaining matter, which are twelve distillations; then pour upon the remaining matter rectified S. V. Cohobate and abstract it six times from the matter; then take it out and grind it, and that it may be the better fixed, calcine it in a crucible in a circulary fire, until it is almost glowing hot. The dose of this powder is gr. vj. with gr. xii. of *Scammony*: Let the patient take it in the morning, the day before the Fit.

Another Febrifuge.

Take Cinaber of Antimony ℥j. common salt decrepitated ℥ij. pulverize them, and mix them together; put them into a glass cucurbite, and pour upon them Oil of Sulphur ℥iij. digest it for two days in a moderate heat in ashes; then augment the fire to evaporate away the humidity, then having edulcorated the remaining mass, reduce it into powder, which mix with ℥iij. of flowers of Sulphur; put this into an Earthen Poringer, which put upon burning coals; let it kindle, and stir it continually with an iron spatula, until all the flowers of Sulphur be burned away: Then pour upon the remaining matter so much S. V. as may cover it the breadth of two fingers, then burn away the S. V., then reduce it to powder, and keep it for use.

This powder is much recommended to cure all sorts of agues and intermitting fevers, giving it half an hour before the fit, from ten to fifteen or twenty grains, in some syrup or cordial-water, taking some broth two hours after it; but the patient should be purged first, and let blood before the use of this powder; and if the first and second dose do not carry away the fit, it must be repeated a third time.

Another Febrifuge,

which is thought to
be Riverius his true Febrifuge.

Take of Mercury dulcis twelve times sublimed $\text{\reflectbox{3}} j$ β. Mercurius Vitae corrected as follows, $\text{\reflectbox{3}}\beta$. mix them together. The correction of Mercurius vitae is thus: Take of ☿ vitae, put it into a small cucurbite, set it in sand in a moderate heat, let it stand until it begins to grow red; then pour upon it rectified spirit of wine, which abstract, and pour on fresh S.V. Repeat this three times, and you shall have a ☿ vitae which will not operate upwards, but only downwards. This ☿ vitae is to be used for delicate persons, but for strong and robust persons you may use the common ☿ vitae.

This powder finding the humours disposed, will operate both upwards and downwards if you employ the common ☿ Vitae; but if you employ the corrected, as was said, it will operate only downwards. And as this Febrifuge contains in it a reasonable dose of ☿ vitae, the ☿ dulcis thus prepared, working for his part upon the bad humours, and serving also for a corrective to the ☿ vitae one ought to expect good effects of it.

Riverius gave this Febrifuge to persons of all ages and sexes, in the morning the day before the fit. One may give six grains of it to little children in the pap of a roasted apple, or in some sweet-meats, and so increase the dose proportionably, according to the age and strength of the patient, to twenty grains to adults, and even to twenty four to those that are of a strong constitution.

(Hartman) These Febrifuges were given me by a friend, a German; and I thought it fit to insert them here: But whilst they were printing, I found them in Mr. Charras his French Dispensatory, which just at that time came to my hand.

A Certain and Experimented Remedy to

Cure the Convulsion Fits in little Children;

as also for the Epilepsie, the Cholick, and for the Spleen, &c.

Take verdigrease, and distill a spirit thereof, which rectify once by itself, and it will leave some feces and metalline terrestreity behind:

Then take one part of this spirit, and three parts of fair water, put it upon lithargy finely searsed, as much as it will dissolve: Deflegm it in Balneo, and then distill it in sand, and there will come over a pure and powerful spirit without acrimony; it will taste a little sweetish, as in the making of *Sacharum Saturni*.

It is excellent for the convulsion of little children, being given in some fit vehicle, a drop or two for sucking infants; but to men you may give ten or twenty drops.

Sigillum Hermetis,

Or, a great and Experimented Medicine, which has done great Effects in the Cure of all sorts of Agues and Fevers. It was given to Sir Kenelm Digby by an able Physician, who had done Wonderful Cures with it.

Take ☽ ℥vj. dissolve it in the best A. F. you can get, using no more A. F. than is necessary for the solution (which will be about ℥j ß i.e. two parts to one) when you see that it is all perfectly dissolved (without fire) cast into the Matrass an *ááá*, made (after the ordinary manner of Goldsmiths) of ℥i. of pure ☉, and ℥ij. of ☿; you will presently see a *pelagus conturbationis* made. Let the Matrass stand still upon a table, or in some corner, till you find the matter at that pass as you desire: you will see many beautiful colours appear. After forty days standing, you will see a kind of roughness appear upon the superficies of the ☿, which will daily grow and sprout out more. In twenty days more (sixty in all) it will be shot out into little spears or needles and twigs. When you see that it grows or shoots out no more, pour off all the liquor, and the Mercurial matter will soon dry of itself.

Then with some little pieces of glass break off
these excrescencies or needles from the mass,
(whereof you may have about $\tilde{3}$j. or more) and grind
them to powder, which will be very white.

Of this powder give twenty four grains, or more
(according to the complexion) in a cherry, or yolk
of an egg, in the morning very early, or at night
going to bed, or rather after the first sleep at
three or four in the morning, and in this last case
sleep after it. It is seven or eight hours before it
starts to work.

Sometimes the first dose will not work at all,
otherwise than by strengthening, and then the author
gives a second dose two or three days after, which
will work either by stool or vomit, or sweat, as nature shall require, and in due proportion.

It cures quartans and other agues, and works
admirably in all desparate diseases. He took it once
a month himself. When there is no peccant humour in
the body, it works not by evacuation, but strengths.
The ☿ encloses and shuts up the metals, like a rose
of Jericho, from whence he calls it *Sigillum
Hermetis*. The part of the needles next the mass
works rougher than the ends. Out of the mass you may
draw most of the gold and silver, with loss of about
an eighth part of the first, and less proportion of

the last. He thinks this is a Philosophical ☿, and to be useful in the great work.

A Mercurial Liquor with Jupiter.

Take lbj. of Jupiter, melt it in a crucible, then pour into it lbj. Of ☿ revived from Cinaber, and made hot, make an *ááá* of it, which wash with warm water, wherein you have dissolved a little salt; wash it so often, till you have washed away all the blackness of it, and the *ááá* will be as white as snow: Then dry it, and grind it in a marble or stone mortar with lbij. of Corrosive sublimate; then spread it upon a large dish of glass, which set on shelving in a cellar, putting something under it to receive the liquor that will run from it, you will find at last the salts resolved into a liquor, in which will be also the ☿, which will be revived; separate the liquor from the ☿, and keep the ☿ for another use: Put the liquor into a cucurbite, and evaporate the superfluous moisture of it in B. M. with a gentle heat: Then digest it for fifteen days more in the same B. with a very gentle heat; then pour this liquor into a retort, which put in sand, and fit a Recipient to it; then distill by graduated fire, giving strong fire at last of the fourth degree; you shall have a liquor like an Oil.

This liquor is much esteemed to cure the cancer, wolf, fistulaes, and all sorts of old, inveterate, malign, and gnawing ulcers, being applied outwardly.

Monsieur C. his Lunary Emetick

and Febrifuge, &c.

Dissolve ☽ in A. F. then precipitate it with spirit of salt, then dry the Calx.

Take of this Calx, and of ☿, *ana* distill it as a butter of ☿, you shall have a butter white and transparent, which will dissolve ☉. It you will make an emetick of this butter, precipitate one part of it with fair water, then edulcorate with blood-warm water, and you will have an emetick remedy, which will purge.

It cures all sorts of agues and fevers, and is a Catholicum for ill humours. The dose is from one grain to three, in some fit thing in the morning fasting. It must be given with great caution.

To make a most Excellent Sudorifick

**of the aforesaid Butter, that will Cure
the Leprosie, and the Venereal Disease,
proceed thus:**

Take the other part of this butter, and put it
into a retort, and pour upon it spirit of Nitre;
distill and cohobate three or four times; then
edulcorate it with fair water, and dry it; then burn
spirit of wine upon it, and you shall have a
sudorifick, which will do admirable effects, taking
from eight grains to sixteen, in the morning in bed;
drink some fit decoction after it: And after the
sweating, the patient must be rubbed with warm
clothes all over his whole body, observing a
reasonable diet, and using some fit purge before.

An Oil of ☉,

**wherewith Monsieur Belieur,
a Famous Chirurgeon at Paris, Cured Cancers, all old
Ulcers, Cankers, and Venereal Sores, &c.**

Take spirit of salt two parts, spirit of Nitre
one part; in this dissolve as much ☉ as it will
dissolve: Distill off very gently the liquor in B.
M. until the ☉ remain in a crystalline gum or salt;

260

then let it resolve to liquor in the air by itself: Then distill again, and resolve. Repeat this until it congeals no more in the cucurbite, but remains a deep-red liquor, like an Oil. The manner of using this Oil is thus: Dip a straw or a feather in it, and touch all around about the borders of the sore with it.

With this he cured a very malignant ulcer in a leg (that had been there above three years) in the space of ten days; and also a cancer in a woman's cheek in fifteen days space, that other chirurgeons (without hope of cure) had given over. With this he also cured a woman (that had seventeen cankers in her private parts, that had been so some years, and without hope of cure) in fifteen days.

Doctor Havervelt his Remedy,

wherewith he Cured the Evil Or Scrofulaes, Cancers, and Old Ulcers.

Take Dantzick Vitriol, calcine it till it becomes yellow, then grind it with salt or salt-petre, the ordinary proportion: With this sublime ☿, which sublime once again by itself; then take only the crystalline part of it, whereof take ʒj. grind it to a subtle powder in a glass mortar, with a glass pestle; put this powder into a large glass bottle, and pour upon it a quart of fountain-water, stop the bottle close, and let it stand thus for some days, shaking it often: Then being well settled and stood without shaking at least twenty four hours, pour off the clear, and filter it. Then take one spoonful of this liquor, which put into a vial, and pour into it two spoonfuls of fair fountain-water: Shake the vial well, then pour it out into a glass, and let the patient drink it in the morning fasting; let him keep himself very warm, and stir and walk as much as he can; but let him neither eat nor drink till two hours after the medicine has operated. It will operate by stools, and by a gentle vomit. The next morning, if the patient finds himself strong enough, let him take the said medicine again, if not, he may rest a day or two between.

With this remedy the author above-mentioned cured all sorts of scrofula's, whether open or shut; the cancer or wolf, whether in the breast, or any other part of the body; as also all sorts of pustula's and old ulcers and wounds.

The said Doctor communicated this remedy to Sir K. D.

Another for the Same.

Sir Kenelm relates, that Dr. Farrar assured him, he had perfectly cured a most contumacious, foul, inveterate evil (several times touched by the King, and wrought upon by the best Chirurgeons, and given over as desperate) by the following means:

Take Garden-Snails, that have white or gray houses upon them, beat them in a mortar with a little parsley, into the consistence of a plaister, which apply to the sore or sores, and change it every twenty tour hours.

This is also good to take away the raging pain of the Gout.

A most Excellent Physical Salt,

as it was prepared in Sir Kenelm's Laboratory.

Take Nitre, Sulphur, *ana* lbj. Camphire $\overline{3}$ij. mingle them well together, and cast them by little and little into an Earthen cucurbite red-hot, which shut close immediately with a just stopper of brick that closes it firmly; the cucurbite must have two arms, unto which are fastened two balloons of glass (as you see by the Figure in the next page) each ballon containing about two quarts of spirit of Urine (to the quantity of ingredients here named)

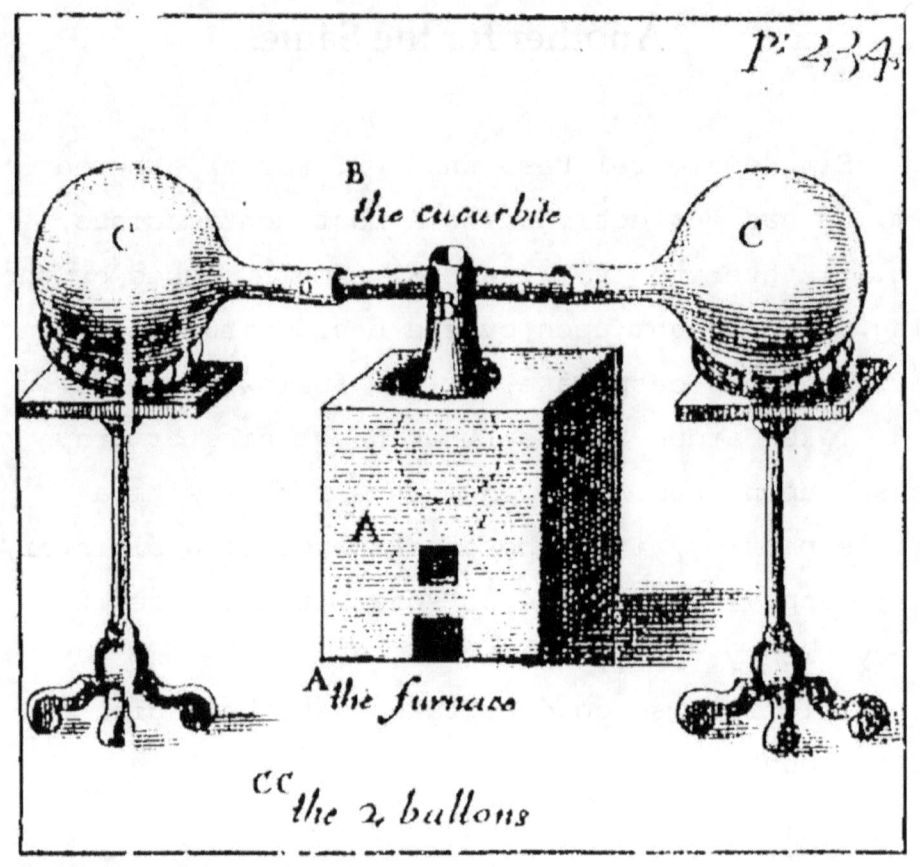

P: 234

B
the cucurbite

C

C

A

A the furnace

CC the 2 ballons

which attracts unto it the spirits, which will ascend and pass by the two arms on each side: When all is cold, take out the fixt matter that remains in the cucurbite, and grind it small, and dissolve it in simple spirit of Urine, and being filtered and congealed, dissolve it in the acid spirit of urine that was in the ballons, and has the spirit of Sulphur, Nitre, and Camphire in it: Distill and cohobate this (in a glass cucurbite) till the salt have retained in it all the spirits that were in the urine. This salt will be very grateful, and not taste or smell at all of the Camphire, nor is Saturnine, or Anti-venereal in its effect. It is very efficacious in all fevers, either simple or malignant any ways, or spotted: In the small-pox or measles, in all the progress of them, from before their coming out, till the end; and preserves the heart from hot and putrid vapours and fumes, and purifies the blood.

The best way to make the Spirit of Urine.

Let the urine stand eight or ten days, in which time it will putrefy and ferment; then distill very gently, and that which comes first is the spirit. When it begins to come weak and insipid (which you will know by tasting a drop) then cease, for all that is good is come over. Thus you shall have near half your quantity of urine in good spirit.

Sir Kenelm Digby's Excellent Remedy

for Tetters, Herps, and Ring-worms, Scabby Itches, &c. as it was prepared by his directions for his own use for a Tetter.

Dissolve ℥ij. of running ☿ in ℥iv. of the best A. F. pour upon the solution a quart of fair water, in which is dissolved two handfuls of salt, and then filtered; this will precipitate the ☿ to the bottom in a white Calx. When it is well settled, pour off the clear water, which keep for use. Pour the remaining milky thick substance upon lbj. of hogs-grease melted in an Earthen pot; the grease must be very hot when you pour in the dissolved ☿; but take the pot from the fire when you pour the ☿ to it, and stir it well all the while you are pouring it in; and when it is all in, set the pot upon the fire again to boil the grease, till all the moisture of the Mercurial substance and solution is evaporated away, but be sure you stir it all the while, as also after you take it from the fire (which you must do as soon as the moisture is gone) till the grease is cold and hardened.

The way of using this ointment and water, to cure all sorts of Tetters, Herps, or any Scabby Itches, or Inflamed red faces or noses, is thus:

First, if the evil is very great, purge and let blood strongly; then begin with the water, rub the tetters, and all about it with linen dipped in the water made as hot as you can endure it; and when you have rubbed and bathed it well, lay upon it compresses wetted in the water. Do thus twice a day for two or three days, or more, till you see it has drawn out the salt humour abundantly, and that the part is much inflamed, and very sore, and has little holes or ulcers eaten in ft. Presently after the first washing it will grow very sore and inflamed; therefore you must not afterwards rub it so hard as at first, but very gently. Some sores will require that you use the water five or six days, others more, tender ones two or three days. When you judge that the water has drawn out sufficiently the violent matter, then anoint with the grease as hot as you can endure it, and lay on it a plaister of the same ointment, binding it on. This ointment will presently assuage the pain, and take away the inflammation. Dress it with it twice or thrice a day. Much matter will run from the sore, as from an ulcer; and by little and little it will heal up. And that which is wonderful is, that whereas one would think that such deep holes as the water will have eaten, should have scars, there will not appear the least mark of them, but a fine new tender natural skin will come over it all.

A great Medicine,

wherewith Wonderful Cures have been performed to my knowledge.

Take Bezoar mineral well prepared, and antimony diaphoretick also well prepared, *ana* ℥j. grind them together to a subtle powder, and put them into a small retort, and pour upon them $\mathrecyclingЗ iv.$ of good spirit of Nitre; distill in sand with a moderate fire to dryness, then cohobate and distill twice, which are three distillations in all with the spirit of Nitre upon the matter. Then pour fresh spirit of Nitre upon it, and distill and cohobate as before. Repeat this a third time with fresh spirit of Nitre, the same quantity as before, which are nine distillations (in all) with $Зxij.$ of spirit of nitre: Then break the retort, and take out the matter, which grind to powder, and edulcorate it well with warm Carduus-water; then dry it gently, and put it into a Poringer, and burn rectified S. V. upon it, stirring it all the while the S. V. burns, with a Silver spoon, until the S. V. is burned all away, and the powder remain dry. Then pour fresh S. V. upon it, and fire it as before: Repeat this a third time, then grind the powder, and keep it in a vial close stopped.

This powder cures the venereal disease the most inveterate, with all its symptoms and attendencies without exception, and restores lost strength and vigor, as experience testifies. It cures all rheumatisms, the leprosie, all interior and exterior ulcers; it purifies the whole mass of blood, and wonderfully fortifies Nature, &c.

The way of using this powder to cure the above-mentioned diseases is thus:

First, purge with some fit quantity of purgative potion; then, if the disease requires you may let blood the next day, then two days after that repeat the purge, and two days after you may begin with the powder; taking gr. viij. of it for five mornings together, the powder being mixt with a little conserve of roses, let the patient take it upon the point of a knife in the morning in his bed, and drink after it immediately a glassful of the following decoction as hot as he can drink it; let him keep his bed, and he will be in a gentle breathing sweat for an hour; which being past, let him be rubbed with warm clothes, his legs, thighs, arms, shoulders, and the back; then let him keep his bed for an hour longer, to see if he will sweat any more: The sweat being quite over, he may rise, and go about his business, as at other times. After those five days the dose of the powder must be increased, taking twelve grains of it for other five mornings; and then you must come again to eight grains for five mornings more. When you begin with the twelve grains after the first five doses, you must drink a little more of the sudorific decoction than before, and taking the powder then in a little confection of alkermes. You may also increase the dose of the powder by degrees (as Sir K. D.

observes) taking (for example) gr. x. the sixth day, and gr. xij. the other three days following; then gr. x. the tenth day, and so come again to gr. viij. the eleventh day. One may take gr. xx. of it at one time without inconveniency. The first purge the author gives, is a decoction of succory and tamarinds, with infusion of two drams of Sena, and being strained, dissolves in it ℥j. of Syrup of peach-flowers. The second is the same, adding only of confection of Hamech, or of confection of Citron; or you may increase a little the dose of Sena, and of the syrup, if one is not willing to take anything where there is Scammony in. The sudorifick decoction the author made use of with this powder, is thus:

Take Sarsparilla ℥ij. China—root ℥j. Sassafras ℥ ß. Santal Citrine ℥ij. and a little Licorise if you will, and a little cinnamon for to aromatise it; let all be infused with three quarts of water for twelve hours in warm sand; then let it boil gently until a third part is consumed, then strain it.

Note, That if you put a little salt of Tartar into the water when you put the ingredients in, it will extract the virtue and tincture out of them much the better; as also in making any purgative decoction, is you infuse them overnight with a little salt of Tartar put in the water, and then boiling it only two or three walms the next morning, it will be much more effectual.

Lapis ignis,

for the renovation of Mankind, by the three Principles of Nature, Salt, Sulphur, and Mercury.

Take ☿ mineral, pulverize it, and calcine it in a close reverberatory, with sufficient, but moderate heat, so that it melt not, in twenty four hours it will be calcined, and will be a gray powder. Take of this ☿ calcined, and of raw ☿ mineral, *ana* lbj. melt them together in a crucible; when they are well melted and incorporated, pour it out into a copper or brass kettle, and it will be glass of ☿, which need not to be clear. If you did not add the ☿mineral, the calcined ☿ would not melt. Pulverize this glass, and grind it upon a marble stone till it becomes an impalpable powder, which put into a vial, and pour upon it distilled vinegar calcified with its fixt salt, digest in sand; when you see the distilled vinegar coloured of a golden colour, decant the clear, and put fresh distilled vinegar upon the glass, and digest as before. Repeat this till you have extracted all the tincture out of the glass: Then filter the tincted distilled vinegar, and put it into a retort; distill with a gentle fire in sand until you see there remain a liquor like a deep-red oil in the bottom of the retort, and that you see some drops appear in the neck of the retort, which is a sign that all the distilled vinegar is come over. Pour upon this Oil Tartarized S. V. digest and circulate for three or four days, or more: Then draw off the S. V. gently in B. and as soon as you see any red drops appear, change the Recipient, putting on another; then distill over all the remaining red Oil to dryness. This Oil will be very red, and very precious, and is the true Oil and Sulphur of Antimony, which is a wonderful medicine against the plague, and all diseases.

To make the Salt of Antimony.

Calcine ☿ in a glass oven, or in a reverberatory, until it becomes perfectly white, without any addition; then sprinkle it with dew, and dry it in the sun; sprinkle and dry it seven or eight times, then grind it to powder: Take of this powder three parts, and one of powder of charcoal; mix them together, and put them into a crucible, which set in a wind-furnace, and give fire by degrees, at last strong fire to make all melt well; then take out the crucible, and knock it against the ground to make the Regulus fall to the bottom; break the crucible being cold, and separate the salt, which you will find between the Regulus and the

Scories. So soon as you perceive that the matter is melted, you must be quick in making the Regulus, and take the crucible out as soon as you can, for fear the salt should evaporate in the fire.

To make the ☿ of ♂ for this Work.

Calcine ♂ in a close reverberatory until it becomes gray, then sublime it in an Earthen vessel; grind again what is sublimed, and sublime it as before. Repeat this operation three times, or until you see the ♂ sublimed hard and ponderous, wherein is inclosed all the ☿ of ♂.

Composition of said Salt, Sulphur, and Mercury.

Take of the said salt ℥j. dissolve it in as much of the Oil as will dissolve it, and as much as the said salt will imbibe, so that it be like an ointment or paste; digest *in fimo equino* for ten days: Then take it out, and add ℥j ß. of the said ☿ of ♂; and being well mixt and incorporated together, put it to digest as before, until it is converted into a red powder. The way of taking this red Powder is thus:

Take gr. iv. of this powder in a little canary sack in the morning fasting in your bed, it will cause a gentle breathing sweat for three days together, during which time you must keep your bed; your chamber must be very warm and close, you may eat and drink moderately of good wholesome food. The three days being past, you may rise, and walk about your chamber, taking good nourishing food, abstaining from all labors in body and mind; and thus you will renew hair and skin, and will be strong and vigorous.

It will not be needful to use this remedy but once in forty years; but you may use of the said Oil, taking three drops of it in a little sack in the morning fasting, for the preservation of your health. This Oil may be given in all distempers with great success. This is from Abbot Boucaud.

The Marchioness de Beck, her Aurum Potabile,

which she much esteems.

Take Calx of ☉, and Regulus of ♂, *ana* ʒj. Jupiter ʒij. melt them together, then grind them to a subtle powder with ʒiv. of sugar-candy, oriental bezoar, and Sal armoniac, *ana* ʒi. Mix all well together, and put them into a large retort, and distill in sand with a graduated fire for six hours;

let the bottom of the retort be red-hot at last for half an hour. You shall have an aureal liquor, whereof two or three drops taken in a little canary, or other convenient liquor, is a great cordial and restorative.

(Hartman) The said Marchioness told me at Paris (where she showed me this Aurum potabile, and gave me the receipt of it) that when at any time she found herself indisposed, she presently took two or three drops of it, and immediately she felt herself strengthened and chearful, &c.

The Baron de Roche showed me also the receipt of it at Paris, who also made great esteem of it, telling me, that he esteemed it to be one of the best Aurum potabile's that could be made, and that it was a Sovereign Cordial and Restorative.

You may reduce two third parts of the ☉ out of the Caput Mortuum, its tincture only, and the subtilest part of it comes over by the distillation.

Cornachinus his Medicinal Powder,

as it was Prepared by Sir K. Digby's Order in his Laboratory.

Take Regulus of ☿, and of pure salt-petre, *ana* ℥iv. mingle them well together in subtle powder, and cast them into a red-hot crucible, and make them burn by casting in a fiery coal, which repeat as often as it consumes; for without that the salt-petre will not burn, there being no more Sulphur in the ☿ to set it on fire. Keep it thus in fusion in a reverberating heat for at least an hour, stirring the matter often with an iron rod; then let it cool.

This must not be edulcorated as common ☿ diaphoretic, but the fixed salt of salt-petre must remain with it, and must by no means be separated from the ☿, for in that Sir K. says, consists the virtue against fevers. Of this we gave with the scammony and cream of Tartar, *ana* gr.x. diminishing and increasing the dose according to age and strength.

(Hartman) Sir K. recommended this to me as a very good purge, telling me, that I might make use of it whensoever I had occasion.

The best way to make the Regulus of ☿, is, to put first into the crucible the salt-petre and

Tartar, and when they are well melted, put in the ♁ , and proceed in the rest in the usual manner: Thus you shall have six or seven pounds, or more, for every pound of Antimony. Likewise to make a Martial Regulus, put the ♁ first into the crucible; and when it is in perfect fusion, then put in the Mars.

A Laxative and Emetick Cream of Tartar.

Take glass of ♁, and Cream of Tartar, *ana* ℥jß. grind them to a subtle powder, then mix them together; put this into a Matrass, and pour upon it lbij. of rosemary-water; digest it for some days, shaking it sometimes; then filter it, and evaporate to dryness, and you shall have a salt, which grind to powder, and keep it in a glass close stopped. It is a safe and gentle vomit, and operates also by stools. The dose is from gr. j. to v. or vj. in a little sack.

Another most Excellent Laxative and Emetick Cream of Tartar.

Take ℥iv. of Cream of Tartar, grind it to a subtle powder, which put into a Matrass, and pour

upon it so much spirit of Sal armoniac as may cover
it the breadth of two fingers; stop it close, and
set it in a cellar for twenty four hours, then pour
it into an Earthen pot glazed, and put into it зj.
of glass of ☿ in subtle powder; set this pot in a
furnace in sand (or upon a gentle coal-fire) and
pour into it a sufficient quantity of fair water;
let it cool six or eight hours, still pouring in
more water as it consumes: At last, evaporate until
it comes to have a thin light skin on the top; then
set it in a cellar, and it will shoot into crystals,
which take out, and keep them for use.

This is a most excellent medicine, and one of
the best emetics that can be prepared. The dose is
from gr.j. to vj. for children; and for aged
persons, from gr. x. to xv. in a little sack.

The best and easiest way to make a most subtle
and penetrant spirit of Sal armoniac, (as it was made
in Sir Kenelm's Laboratory, and as I make it now) is
thus:

Take Quick-lime grosly beaten, put a bed of it
about two fingers thick into the bottom of a
cucurbite; then dissolve lbj. of Sal armoniac in as
much water as will dissolve it: Pour of this
dissolution upon the Quick-lime (having first placed
the cucurbite in the furnace in sand) so much as may
dissolve it well, and swim half a fingers breadth
over it. Then be as quick as you can in fitting on

the head and Recipient (for it will immediately begin to distill without fire) lute all the junctures well, and distill with a gentle fire, keeping the subtle spirit by itself, which comes first. If any flegm should come over with the second spirit, rectify it in Balneo.

This spirit is not only good for benumbed heads to smell to, but also to take inwardly, for it is a most excellent remedy: It opens all obstructions, it is sudorifick and diuretick. 'Tis very good in fevers, especially the putrid, in Palsies, Epilepsies, Hysterical Fits, and the Plague, resisting all corruption, in Lethargy, and stupification of the Spirits. The dose is from eight to thirty drops. It also assuages the pain of the Gout, being mixed with S. V. or Brandy, and Linnen Clothes dipt in it, and laid upon the parts afflicted.

(Hartman) In distilling of this Spirit this way, I have observed several inconveniences; the first is, that if you use a glass cucurbite, it will be apt to crack by the sudden heat of the Quick-lime, caused by pouring on the dissolution of the ✳ (and an Earthen cucurbite will imbibe it.) Secondly, by the same reason you lose a great part of the subtilest spirit, which will ascend before you have poured in the quantity of the dissolution above-mentioned, and before you can fit on the head and

Recipient. To prevent all these inconveniences, I make use of a tin cucurbite, with a spout in the upper part (See it in the third Figure) and having placed it in the furnace in sand, I put in the Quick-lime, and then I fit a glass head and Recipient, and having well luted all the junctures with wet bladders, or paste and paper, I pour in the dissolution of ✳ by a funnel through the spout; then I close the spout, and distill with a gentle fire. When the distillation is over, take out the Caput Mortuum, and make the cucurbite clean, and wipe it dry, that it may not rust or canker, and it will serve for many other operations, and will save you the buying of many glass cucurbites, which by reason of their thick and knobbed bottoms are so apt to break.

The Volatile Salt of Tartar,

as I have often made it, which is an Excellent Remedy.

Take Lees of Wine, (which you may have from the wine-coopers when they have pressed them out for making of vinegar) break them in small pieces, and let them dry; then being very hard and dry, bruise them grosly, and fill an Earthen retort with it, or a glass one coated; distill in naked fire, fitting any Recipient to it to receive only a sour flegm, which will come over first; as soon as you see any white fumes come over, (among which comes the volatile salt) change the Recipient, putting on another pretty large; lute the junctures well with paste and paper, then increase the fire by degrees, until you see the Recipient filled with white fumes; continue the fire in that degree, until those white vapours diminish, and that the Recipient begins to grow cold: Then augment the fire to the highest degree, to force all over at last; when nothing more comes over, cease. The distillation will be performed in three or four hours; you will have a whitish liquor, which contained in it the volatile salt, and part thereof will stick to the sides of the Recipient, and a reddish foetide Oil will swim upon the liquor. Pour out all the liquor that is in the Recipient, then pour a little warm water in the Recipient, and shake it to get out all the volatile salt: Separate the Oil from the liquor by a glass funnel; then filter the liquor, to free it from all Oiliness: Put this liquor into a Matrass with a long neck, to which fit a head and a small Recipient; distill in sand with a very gentle heat, and the volatile salt will ascend into the head as white as snow; when you see that a pretty quantity is sublimed, take off the head, and stop the mouth of the Matrass, if you have not another head to put on;

be as quick as you can to gather the volatile salt that is in the head, and put it into a vial, which stop very close with a glass stopper, for it is very apt to resolve into liquor when it takes air: Then put the head on again, and continue the sublimation until there sublimes no more salt; gather this last salt, and put it to the rest: Then put on the head again, and augment the fire a little, and you shall have a fiery liquor, which is the volatile salt, mixt with some flegm, which makes it come over in a liquid form.

This salt is much esteemed and recommended to purify the blood by sweat and by urine. It is the best of all common remedies against hysterical fits and vapours, smelling to it, and taking it inwardly. It is excellent against the palsie, apoplexy, epilepsie, &c. against quartan and tertian agues. It opens all obstructions, and provokes the Terms. The volatile spirit has the same virtue as the salt; it is good for all obstructions, particularly of the spleen, and keeps the body open; it is far beyond the common spirit of Tartar in virtue. The dose is from eight to twenty or thirty drops in some fit vehicle.

A Physical Salt.

Take Nitre and Oil of Sulphur, *ana* lbj. flegm of vitriol lb ß. pulverize the Nitre and put it into a retort, and pour upon it the Oil of Sulphur and flegm of vitriol; distill in sand, and you shall have a very good spirit of Nitre, and a pure white salt will remain in the bottom of the retort. This salt is much esteemed in fevers, and to quench thirst, being taken in juleps, ptisans, or possets. The dose is thirty or forty grains.

A Precious Tincture of the Flowers of Antimony.

Take the dark-red flowers of ♁, digest and circulate them with rectified spirit of vitriol; when they are sufficiently united, abstract the spirit of vitriol to an Oil, upon which pour S. V. digest and extract a tincture s. a. which abstract again to the consistence of an Oil. This tincture fortifies and cherishes the heart & vital spirits, strengthens the stomach, is good against agues and fevers, hysterical fits, hypochondriac melancholy: It cures the jaundies, opens obstructions, provokes the Terms. It is good against the gout, scurvy, and dropsie, itch and scabs: It purifies the blood, and strengthens nature. Dose from gr.j. to iij. or iv. given in a fit vehicle.

An Excellent and true Tincture of Coral.

Take good red Coral $\mathfrak{Z}$iv. grind it to subtle powder, which mingle with $\mathfrak{Z}$iv. of Sal armoniac that has been three times sublimed from decrepitated salt. Put this mixture into a small cucurbite, which set in a sand furnace; fit a head and Recipient to it, and having well luted the junctures, give a gentle fire at first, which augment by degrees. There will come over first, a small quantity of a urinous volatile spirit; after that, you shall see the flowers ascend and stick to the head, and upper part of the cucurbite. These flowers will be tinged with divers colours, as red, green, blue, very pleasant to behold, they contain in them the true tincture of Coral; for the body of the Coral which remains in the bottom, will be as white as snow; continue a moderate fire until no more flowers ascend: The operation will be performed in a few hours. Then gather diligently all these flowers, and put them in a Matrass, and pour upon them rectified S. V. to the height of four fingers, which will remain white in the bottom; filter this tincture, and abstract from it three fourth parts of the S. V.

and a deep-red tincture will remain in the bottom, which is the true tincture of Coral.

This tincture is a sovereign remedy to strengthen the stomach and bowels: It purifies the blood by sweat and urine. It opens obstructions, is excellent in all sorts of fluxes, &c. Dose from six to twenty four drops, in some convenient vehicle.

The way to sublime the flowers of Sal armoniac for this work is thus:

Take common salt decrepitated and Sal armoniac *ana* lbj. pulverize and mingle them together, and put them in a cucurbite, and sublime in sand with a gentle fire at first, which augment by degrees; the flowers will ascend into the head like meal: Continue the fire for five or six hours; then let all cool, and gather the flowers, which mix with new salt, and sublime as before: Repeat this three times.

An Excellent Extract of Mars,

for Diarrhaea's and Fluxes.

Take filings of steel (which you may buy at the needle-makers) $\mathfrak{Z}$iv. put them in well-glazed Pipkin, and pour thereon a quart of good deep-red wine, (that which is used to colour white-wine) let it boil until about three parts of the wine is consumed, stirring often with an iron spatula. Then strain it whilst it is hot.

It is a great and certain remedy for dysenteries, Diarrhaeas, old hepatical fluxes, and such like diseases; you may give an ounce of it in broth fasting, for some mornings together. This I have sufficiently experienced with happy success.

Sir Kenelm Digby's Remedy for the same;

as it was prepared by his Order, and much used.

Take Rye-flower, and make a paste thereof with juice of Elder-berries; then bake hard biskets thereof, which pulverize, and make a paste again with the juice of Elder-berries as before: Repeat this three times. Then reduce it to powder. The dose is one dram.

Sir Kenelm Digby, his Excellent Plaister

of Lead.

Take of the best Oil Olive lbij. $\overline{3}$iv. white-lead, red minium, *ana* lbj. grind them to powder, and put them with the Oil into a large glazed pot or pipkin, with $\overline{3}$xij. of Venice soap shred small, which put upon a gentle coal-fire, and stir it well with an iron Spatula for an hour; then increase the fire a little, which continue until the liquor is of the colour of an Oil: Then drop some of it upon a board, and if it sticks, or that it cleaves to your fingers, 'tis a sign that it is boiled enough; then roll it up, and keep it for use.

This plaister being applied to the stomach, is good for the weakness and indigestion thereof, and causes a good appetite.

Being applied to the belly, it cures the collick; and being applied to the back, it strengthens the reins, cures the bloody-flux, the Gonorrhaea and tempers the excessive heat of the liver.

It cures all contusions and bruises, swellings and inflammations. It maturates and draws all sorts of apostumes, wolfs, and pustles, and cures them, without lancing or incision. Being applied to the head, it strengthens the eye-sight: To the fundament, it cures all accidents that may happen there, as piles, &c. And being applied to the belly of a woman, it provokes the Terms, and disposes her for conception.

Dr. Stephen's Plaister for the Gout.

Take virgins-wax lbij. Hogs-grease ℥ß. Mutten-suet ℥ij. Oil of Colts-foot, plantain and rose-water, *ana* ℥ij. lavender-water ℥ij. dragon-water, and water of borage, *ana* ℥ ß. two nutmegs, two cloves, and a little mace, beat into powder; mix them all well together, and let it boil with a moderate fire unto the consistence of an ointment; wherewith anoint the part grieved as hot as you can endure it, and dip linen clothes in it, and apply them.

A very good Ointment for the Gout;

and the running Gout, Aches, Numbness, and Pain in the Joints, &c.

Take the tender branches of dwarf-elder, in the month of March, when they sprout out of the ground from the root, and are not above a finger long, three handfuls; stamp them in a stone mortar, then mix them with lbj. of hogs-grease; put this into a Pipkin, and let it stew upon a gentle fire for two or three hours.

This was communicated unto me by a worthy gentleman, who much esteemed it, because he found great benefit by it in the Gout: It takes away the raging pain thereof, gives ease, and strengthens the parts afflicted.

In the running Gout, numbness, and raging pain in the joints, I have had much experience of the virtue of this ointment, after many remedies used in vain; the parts grieved must be annointed with it as hot as can be endured, and chased in before a fire.

A Certain and Infallible Remedy

to prevent and Cure the Fits of the Gout.

I knew a gentleman in Germany, who always cured and prevented his fits of the gout (whensoever he perceived the least symptom of its approaching) by the following remedy:

He caused a good quantity of the herb Mullein (*Verbascum* in Latin) to be gathered every summer when it was in its flower, which is in June, it bears many yellow flowers upon a long straight stalk with large leaves at the bottom, which are hoary. Then he took a good quantity of this herb, and cut it small, the stalk, flowers, and leaves, and caused it to be boiled in a pail-full of forge-water out of a smith's trough, wherein he quenches and cools his irons; when this was boiled sufficiently, then there was put into it a large piece of chalk in powder. In this water he bathed his feet, legs, and knees, as hot as he could endure it, in a tub, continuing until the water grew cold. Then a hole was dug in the ground in his garden, wherein this water was put with the ingredients, and then covered with earth.

This always prevented his fit, so that he never had any pain, lameness, or swelling at all, to which I was an eye-witness. And I heard him say, that if he did not use this remedy, he would have very shrewd and racking fits, and keep his bed by it for a month or six weeks, and that twice a year, chiefly in the spring, and at the falling of the leaf.

Mr. Locher, an Apothecary of London, his

Excellent Oil for Deafness,

which he gave to Sir K.D.

Take Oil of bitter almonds, Oil of spikenard, *ana* ℥vj. juice of onions, juice of rue, *ana* ℥ij. black hellebore Ɔß. Colloquintida ℥ß., Oil of Exeter ℥ij. Boil this till the juice is consumed; then strain it, and add two drops of Oil of aniseed, Oil of Origanum one drop. Pour a drop or two of this Oil into the ear, and lie upon your bed with that ear upwards that you intend to drop into, lie still for a quarter of an hour after; then drop into the other, if it is required. It is to be continued a month, or two or three, as you find benefit. When you have dropt into the ear, you must stop it with a little black-wool, dipped in the Oil. Many persons have found much benefit by the use of this Oil, to my knowledge.

Another Experimented Remedy for the same.

Take a good large eel, slay it, and cut it into round pieces of the length of a finger, stick them full with rosemary and sage; then take an Earthen pan, put two or three sticks of wood in cross-wise, lay your pieces of eel upon them, so that they may not touch the bottom of the pan; bake it in an oven as you do meat: Then take the fat of the eel that is in the pan, and strain it through a fine linen cloth, measure how much there is of it, and put to it as much S.V. Then take juice of onions, and juice of the white ends of leeks, *ana* ℥j. of your first mixture ℥ij. put them together into a vial, stop it close, and shake it well for an hour. It is in all things to be used as the former, except that instead of one or two drops, you must drop in three or four.

(Hartman) This was communicated to me by a gentleman at Paris, who had done wonderful cures with it, and among the rest, he had cured the governour of Calais his secretary with it, who had been deaf twenty years, his deafness being caused by a sickness.

A most Precious Balsam of great Virtue.

Take turpentine lb ij. Lignum Aloes $\text{3}\beta$. Mastick, Cloves, Galingal, Cinnamon, Zedoar, Nutmegs, Cubebs, Olibanum, *ana* 3j. Roots of Master-wort, of Angelica, *ana* $\text{3}\beta$. Figs cut small six in number, Gum Tragacanth 3ij. Bruise all the ingredients, and mix them well together, then put them into a glass retort, and having warmed the turpentine to make it run, pour it upon the ingredients, and distill in sand:

Separate the Balsam from a little flegm that will come over with it.

1. This Balsam is a very great preserver of the health of mankind, taking every morning three or four drops of it in a little beer or wine; it strengthens the stomach, and causes a good digestion, and a good appetite.

2. It strengthens the brain and memory.

3. It is good against deafness, pouring two or three drops every day into the ear, and stopping the ear

with a little black-wool, moistened with a little of this Balsam.

4. It helps rheumatic eyes, takes away the heat and pain thereof, and strengthens the sight, anointing the eye-lids therewith.

5. It cures all sorts of scabs, itches, and scall'd heads, be they never so bad.

6. It cures fistulaes, the cancer, wolf, and all other gnawing diseases; and cures all sorts of wounds, whether old or new.

7. It cures the Gonorrhaea, the whites in women, and strengthens the reins.

8. It is good against the biting of a mad dog, vipers, and other venomous beasts, being both inwardly and outwardly applied; and is a great preservative against the plague.

9. It is very good against the cramp, numbness, aching, and pain in the joints, contraction and weakness of the nerves coming from a cold cause, as experience testifies.

10. It sweetens an unsavory or stinking breath, and suffers no worms to breed in the stomach and bowels.

11. It is said, that if a dead corpse be embalmed with it, it will never rot nor consume, nor any linen about it that is imbibed with Balsam: And that for a trial, one should take a piece of flesh, and warm it well against the fire, then rub it over with this Balsam, and let it be well imbibed with it, rubbing it with it three or four times. Then lay it away, and it will remain sound and fresh, so that it may be eaten a twelve month after.

Laudanum Germanicum:

Being A singular Preparation of Matthew's, or Dr. Starky's Pills.

I thought I could not better finish this book, than with the receipt of these most excellent pills, with the true way of preparing them, which far exceeds the common: The receipt is thus:

Take Opium lbj. dissolve it in distilled vinegar, then filter and evaporate to the consistence of a mass for pills: Then take black Hellebore lbj. reduce it to a subtle powder, which put into a Matrass, and pour upon it so much distilled vinegar as will cover it the breadth of four fingers; digest for two days, then evaporate with a gentle heat to the consistence of pills. Then take of the Corrector lbj. Oil of Amber that has been rectified with fair water, Z ij. Licorise dryed and reduced to subtle powder, lbj. Saffron dryed and pulverized, lb β. Put all into a large mortar (well warmed by putting coals kindled into it) incorporate them well together by strongly beating and mixing them, adding by little and little (as you incorporate them) of the Oil of Turpentine that has stood upon the Corrector, and is of a red colour, Z iij. Tincture of Antimony Ziv. Oil of Aniseed, of Juniper-berries, of Sassafras, Oil of Vitriol, Spirit of harts-horn, *ana* Zij. gum Arabick dissolved in distilled vinegar, $\text{Z}\beta$., and if you see that the composition is too stiff, add a little more of the said Oil of Turpentine, and of the tincture of Antimony: Then put it up in a gally-pot, and tie it up close with a bladder and leather.

The composition of these pills is of a very fine consistence, and not so crumbly as the common, but commodious to handle, and make up in pills like unto warm wax. The dose is two small pills about the bigness of an ordinary pea, or one pill about the bigness of a gray pea swelled, taking them at night.

These pills are approved of, and are prescribed, and used by the best physicians, in consumptions, and in other cases.

I thought to have reserved the preparation of them to myself, and not to have published it; but thinking that it is unchristian to keep anything from the public good, my conscience would not permit me.

The preparation of the Corrector differs not from that of Starky's; but because this book may come to the hands of some persons which do not know it, I thought good to insert it here.

Take pure salt-petre, and white-wine, or rhenish-wine Tartar, *ana* equal parts, pulverize them, and searse them, and mix them well together: Then take a large crucible, and set it in your furnace, and being red-hot, cast in some of your mixture by little and little with an iron ladle, and when the fulmination is over, cast in more, which continue till you have put in all your mixture; then let it flow in the crucible, giving strong heat.

Then pour it out, and when the crucible is cold, scrape off all the salt that sticks to the sides of it. Dissolve this salt in boiling-water. Make likewise a Lixivium of Quick-lime and water, which being settled, pour it off: Take of this Lixivium the same quantity with that of the salt of Tartar; mix and filter them, then evaporate to a salt, which will be pure, clear and white like crystal; grind it to powder, and put it into a strong large vessel, and pour upon it immediately so much Oil of Turpentine as may cover it the breadth of four or five fingers; stir it well together, then cover it loosely, only to keep things from falling in, and that the air may come to it; let it stand thus, stirring it three or four times a day with a wooden Spatula, and as you see the salt imbibes the Oil, add still more Oil, until the salt has taken in and absorbed three times its weight of Oil, or that it will take in no more, and is like a soap, and the Oil that swims upon it is of a red colour.

The Tincture of Antimony is made thus,

according to Basil Valentine.

Take equal parts of salt of Tartar and ☿, melt them together in a crucible, keep them in fusion for half an hour, then pour it out, and whilst it is hot reduce it to powder, which put into a Matrass, and pour upon it of the best rectified spirit of wine so much as may cover it the breadth of three fingers, set the Matrass in warm sand, that the S. V. may boil a little, and you shall have a very red tincture, which decant, and keep for use.

This tincture is recommended to open all obstructions, of all the principal parts, as liver, spleen, lungs, womb, reins, and bladder; it provoked the Terms, cures the yellow jaundise, green-sickness, scurvy, dropsie, asthma, pleurisie, melancholy, ulcers inward and outward, scabs, itch, pox, small-pox, and measles. Dose gr. iv. to xij.

Postscript

The Preparation of Sir Kenelm Digby's Sympathetical Powder, as we prepared it every Year in his Laboratory, and as I prepare it now, is only thus:

Take what quantity you please of good English vitriol, dissolve it in warm water, but use no more water than will dissolve it, leaving some of the impurest part at the bottom undissolved: Then filter the dissolution, and evaporate it until you see a thin skin upon it, then put it in a cool place, and let it stand without stirring it for two or three days, covering it loosely only, to keep things from falling in. It will shoot into fair, green, and large crystals, which take out, and spread them abroad in a large flat Earthen dish, and expose them to the heat of the sun in the dog days, turning them often, and the sun will calcine them white; when you see them all white without, beat them grosly, and expose them again to the sun, securing them from rain; when they are well calcined, powder them finely, and expose this powder again to the sun, turning and stirring ft often. Continue this until it is reduced to a white powder, which put up in a glass, and tie it up close, and keep it in a dry place.

As for the virtues of this powder, I will only say, that I have seen great experience of it in my time, in stanching of desperate bleeding at the nose. 2. In stanching the blood of a wound. 3. In curing with it any green wound (where there is no fracture of bones) without any plaister or ointment, in a few days.

A girl about twelve years of age bleeding desperately at the nose for two or three days together, her mother having used all the means she could devise (in vain) came to me, telling me, that she had heard I had a powder that would stanch bleeding, she desired me to let her have a little of it, for she feared her daughter would bleed to death: I gave her some of the powder, and bid her put a little of it in three or four spoonfuls of fair water, and to bath her nostrils with it with a clean linen rag, putting it up into the nostrils, which she did, and her bleeding stopped immediately; the next day she did bleed a little again, and then using it again, it did stanch it, and she never bled again afterwards.

I spoke with a famous chirurgeon, named Mr. Smith, in the city of Augusta, Germany, who told me, that be had a great respect for Sir K. D. books, and that he made his sympathetical powder every year, and did all his chiefest cures with it in green wounds, with much greater ease to the patient than if be had used ointments or plaisters.

If the reader desires to know more of the effects of this powder, and the reason of it, I refer him to the reading of Sir K. Digby's Treatise of Curing of Wounds by way of Sympathy, where he will find entire satisfaction and full information of the reasons of its effects.

-FINIS-

A Word from the Publisher

Thank you for purchasing this book from The R.A.M.S. Library of Alchemy. During his lifetime, Hans Nintzel dedicated himself to the identification, acquisition, study, retyping and, when necessary, translation of what he considered to be the most important known works on Alchemy. Hans was assisted by his sparse network of fellow Alchemists, all members of the Restorers of Alchemical Manuscripts Society (R.A.M.S.). I was an active member of R.A.M.S.

Hans provided copies of the R.A.M.S. works as photocopies. My goal is to publish all of them as professionally printed books.

The works from the original R.A.M.S. Library are republished by R.A.M.S. Publishing Company in the collection, "The R.A.M.S. Library of Alchemy," with permission of the Estate of Hans W. Nintzel.

If you have a work on Alchemy that you believe should be a part of the R.A.M.S. Library, please contact me through R.A.M.S. Publishing Company.

Philip N. Wheeler

The R.A.M.S. Library of Alchemy

The study and practice of Alchemy was extremely important to Hans W. Nintzel. He assembled this Library over a period spanning more than three decades, guided by his teacher Frater Albertus. The R.A.M.S. Library of Alchemy includes all of the most valuable Alchemical texts that Hans painstakingly located, acquired, retyped, and translated during his lifetime, with help from other R.A.M.S. members.

The following is a list of the volumes that are currently available. Volumes that contain works from multiple authors may have only the principle author or editor listed. Additional volumes are forthcoming.

Volume	Title	Author or Editor
1	Twelve Keys of Basilius Valentinus	Basilius Valentinus
2	Triumphal Chariot of Antimony	Basilius Valentinus
3	His Secret Book	Artephius
4	The Golden Work	Hermes Trismegistus
5	Three Works of Ripley	George Ripley
6	Four Works of Paracelsus	Paracelsus
7	Bacstrom's Notebooks, Part 1	Sigismund Bacstrom
8	Bacstrom's Notebooks, Part 2	Sigismund Bacstrom
9	Summa Perfectionis	Geber (Abu Musa Jabir ibn Hayyan)
10	The Five Centuries	Rudolph Glauber
11	The Greater and Lesser Edifyer	Johann Grashoff
12	Chemical Secrets and Experiments	Sir Kenelm Digby
13	The Turba Philosophorum	Arisleus
14	Das Aceton	Christian Becker
15	TBD	
16	TBD	
17	TBD	These volumes are reserved for the Works of Glauber.
18	TBD	
19	TBD	

20	TBD	
21	Alchemical Symbols, Third Edition	Hans W. Nintzel and Philip N. Wheeler
22	The Book of Formulas	John Hazelrigg
23	18 Short Tracts	Hans W. Nintzel
24	Bacstrom's Notebooks, Part 3	Sigismund Bacstrom
25	A Discourse on Fire and Salt	Blaise Vignere
26	The Mineral Work	Johan Hollandus
27	The Vegetable Work	Johan Hollandus
28	Lamspring's Process	Lamspring
29	The Book of Abraham the Jew	Abraham Eleazar
30	Five Short Works of Glauber	Johann Glauber
31	The Metamorphosis of the Planets	Johannes Monte-Snyder
32	Four Works of Roger Bacon	Roger Bacon
33	The Golden Chain of Homer	Homerus, Kirchweger, Nintzel, Wheeler
34	Alchemy Rediscovered and Restored	Archibald Cochren
35	Aurifontina Chymica	John Houpreght
36	The Golden Fleece	Salomon Trismosin
37	The Transmutation of Base Metals into Gold and Silver	David Beuther
38	Sanguis Naturae	Christopher Grummet
39	A Revelation of the Secret Spirit	Giovanni Lambi
40	The Holy Guide, Part 1	John Heydon
41	The Holy Guide, Part 2	John Heydon
42	Secreta Alchymiae	Kalid Persica
43	The Golden Treatise of Hermes	Hermes Trismegistus
44	Potpourri of Alchemy, Part 1	Hans W. Nintzel
45	Potpourri of Alchemy, Part 2	Hans W. Nintzel
46	TBD	
47	Selected Chemical Universal and Particular Processes	Alexius von Ruesenstein

www.ingramcontent.com/pod-product-compliance
Lightning Source LLC
Chambersburg PA
CBHW080650190526
45169CB00006B/2061